Reaction Kinetics Based on Time-Energy Uncertainty Principle

Satoru Yamamoto

Reaction Kinetics Based on Time-Energy Uncertainty Principle

Springer

Satoru Yamamoto
Kyoto, Japan

Translated by
Teruo Tanabe
Kyoto, Japan

Yoji Imai
Tsukuba, Japan

Kenzo Hanawa
Ichihara, Japan

Hideo Yoshida
Nagoya, Japan

Mahoto Takeda
Yokohama, Japan

ISBN 978-981-19-9672-6 ISBN 978-981-19-9673-3 (eBook)
https://doi.org/10.1007/978-981-19-9673-3

Translation from the Japanese language edition: "Atarashii Hannosokudoron no Kokoromi" by Satoru Yamamoto, © The author 1979. Published by Showado. All Rights Reserved.

This Springer imprint is published by the registered company Springer Nature Singapore Pte Ltd.
The registered company address is: 152 Beach Road, #21-01/04 Gateway East, Singapore 189721, Singapore

Translators' Preface

To illustrate Arrhenius' empirical formula for rate of chemical reactions, several theories including collision-reaction models have been proposed, but the most currently accepted of these is Eyring's so-called absolute reaction kinetics. We ourselves tried to understand the content when we were students, but we were not completely convinced.

Eyring hypothesized a transition state in order to theoretically derive Arrhenius equation. That is, suppose the existence of an activated complex, $[AB]^{\ddagger}$, in a transition state when A and B react to produce C, and consider the following process:

$A + B \rightleftarrows [AB]^{\ddagger} \rightarrow C.$

The assumptions are as follows:

(1) A, B, and $[AB]^{\ddagger}$ are assumed in pseudo-equilibrium even when the reactants (A, B) and products (C) are not in equilibrium with each other.
(2) $[AB]^{\ddagger}$ reacts irreversibly to become C.

It is assumed that A, B, and $[AB]^{\ddagger}$behaves in classical mechanical manner, and the reaction does not proceed unless the reactants overcome the energy barrier of the transition state, which is the state of energy maximum along the reaction path, by thermal activation. Although various methods have been devised to create model structures of the activated complex, it is imaginary or hypothetical and is not subject to observation, which is different from reaction intermediates with definite lifetime.

In 1979, S. Yamamoto wrote this book which stated that Arrhenius formula can be explained by using the uncertainty principle.

The uncertainty principle is expressed by

$\Delta p \cdot \Delta x \sim h/2\pi$

$\Delta E \cdot \Delta t \sim h/2\pi$

However, there is still controversy over what is the physical meaning of ΔE and Δt [1, 2].

Yamamoto proposed the interpretation that ΔE is the fluctuation of the energy of the system, and that Δt represents the lifetime of that system.

If ΔE is zero and the system has a definite energy value, the lifetime of the system, Δt, is infinite; that means the system is steady and stable. If the system in question is described by the superposition of two or more quantum mechanical eigenstates, then ΔE has a nonzero finite value, which makes Δt also has a nonzero value (finite lifetime). This interpretation was applied not only to so-called chemical reactions but also to changes of states such as phase transformation.

When this book was published in Japan, this interpretation has sparked much controversy in the Japan materials science community. However, without final conclusions reached, the debate subsided perhaps because the materials science community in Japan was piling up practical demands such as the development of new exotic materials, and there was no time to thoroughly discuss basic issues at that time. However, the issues proposed by Yamamoto still have the importance, and we thought that it is important to make his theory widely known to the world.

According to his theory, the reacting system is in a quantum-mechanically uncertain state, and the lifetime of the state is determined by the magnitude of energy fluctuation. This principle can be applied not only to thermal effect by increasing temperature of the system but also to other effects of such as light irradiation and application of magnetic field. It should be noted that not only so-called chemical reaction but also physical changes of materials including solidification, phase transformation in metals and alloys, crystal growth, etc. can fall within the scope of the theory since they are also reflecting the changes in chemical bonds.

In the end of this preface, we must state that "Abstract" and "key words" for each chapter are written or selected by translators for ease of readers' understanding. Furthermore, references are listed at the end of each chapter in the original Japanese version, but many of them are not directly cited. Therefore, in this English translation, the list is given in the end of each chapter as "Further readings" without citation numbers, but those references where the citation is obvious are written as "[3]," "[3, p.3]," or "[3, Chapter 5]".

Kyoto, Japan — Dr. Teruo Tanabe
Nagoya, Japan — Dr. Hideo Yoshida
Tsukuba, Japan — Dr. Yoji Imai
Yokohama, Japan — Dr. Mahoto Takeda
Ichihara, Japan — Dr. Kenzo Hanawa
August 2022

References

1. J. Hilgevoord, The uncertainty principle for energy and time. Am. J. Phys. **64**, 1451–1456 (1996). https://doi.org/10.1119/1.18410
2. J. Hilgevoord, The uncertainty principle for energy and time. II. Am. J. Phys. **66**, 396–402 (1998). https://doi.org/10.1119/1.18880

Preface to the English Version

I know a variety of rate equations have been proposed in the past. I have recognized those equations had not fit with experimental data especially by using electrical resistivity which is continuously changed with time. Therefore, an experiment was chosen that could precisely measure the time variation and examine in detail whether it fits the kinetic theory. It turned out that the rate equations of the past did not agree with the experimental data.

Consequently, I recognized we must reconsider in a fundamental way. In the past, I have noticed that people make up changes in things only in their minds. Let us assume that this thinking as human-being-centered principle. In contrast to this principle, countermethod of thinking would be called material-centered principle. In the following, I would explain this way of thinking more precisely.

Material-centered principle is the standpoint of view that is willing to recognize the nature as it is. The most important problem is how to answer what is correct. In the standpoint of material-centered principle, correct or incorrect would be determined whether the predictions are consistent with experimental data. It would be helpful to understand material-centered principle by presenting one more example of material science.

In material science a concept of "vacancy" is often used. I found this concept of atomic vacancy problematic upon careful consideration. Water quenching from solution treatment temperatures would have no effect on the number of atoms in the material. Then the difference between furnace-cooled and water-quenched would be the spacing between the atoms.

Then, comparing the water-quenched and furnace-cooled cases, the spacing between atoms would be larger in the water-quenched case. Therefore, in the classical human view, the larger the space between atoms, the greater the diffusion. However, bcc metals with large space per unit cell, such as Cr, Mo, and W, diffuse more slowly than fcc metals with the densest structures with small space, such as Cu and Al, even though the space are larger. Cu and Al metals, which have smaller space between atoms, diffuse more easily. This is a good example of how and what we think in our minds does not match what is actually occurring in matter.

Under the above consideration, the way of thinking of material-centered principle turns out to be accordance to quantum mechanics. In quantum mechanics, there is an uncertainty principle of time–energy, where the wave function deals with the properties of particles and the properties of waves at the same time, and electrons can only be described probabilistically. Since quantum mechanics is a theoretical system that uses these concepts, it is better to assume that physics is quantum mechanics and philosophy is materialism, which greatly expands our view of the material. This viewpoint will help to greatly expand material science.

Kyoto, Japan
January 2022

Satoru Yamamoto

Preface to the Original Japanese Version

The most important thing for a researcher is to have the right attitude toward research. What kind of research results a researcher achieves depends on the objective situation in which he or she finds himself or herself and on his or her attitude toward that situation. Of particular importance in objective situations are past research results and current dominant ideas. We are easily dominated by past research results and current dominant ideas. We cannot expect progress and development of research from a blind attitude that is easily influenced by existing authoritative results and ideas. A critical spirit that clarifies the limitations and difficulties of existing results is necessary for the progress and development of research. To criticize does not mean to unilaterally destroy and abandon past achievements by pointing out their limitations and difficulties. It is to identify what should be inherited from the past and what should not. This critical inheritance and development of past results is the proper attitude for researchers to take.

In this book, we attempted to critically inherit and develop the past results of reaction kinetics. In Chap. 1, the limitations and difficulties of the past reaction kinetics are clarified, and a viewpoint for overcoming them is established. In Chap. 2, we analyze the limitations and difficulties of current reaction kinetics. A new attempt to overcome the fundamental difficulties of the current reaction kinetics and to take a step forward is developed from Chaps. 3 to 6.

Chapter 1 describes a scientific and epistemological nature of our reaction kinetics. This book emphasizes the historical and hierarchical viewpoints, and from these viewpoints, the logic of reaction kinetics and its background in the natural sciences, especially classical mechanics, quantum mechanics, thermodynamics, and statistical physics, and their interrelationships, are discussed. Therefore, the contents of Chap. 1 will be of interest not only to readers interested in reaction kinetics, but also to readers who are interested in philosophy of science and epistemology.

In Chap. 2, the fundamental difficulties of the current reaction kinetics including nucleation-growth theory in phase transformations are clarified from the viewpoint described in Chap. 1. The reader will thus realize how fundamental difficulties exist in the current reaction kinetics, which at first glance appears to be a complete system.

In Chaps. 3 and 4, a new attempt to overcome the fundamental difficulties of the current reaction kinetics is developed. It is based on the uncertainty relation between time and energy. In consideration of the readers who are not familiar with such a concept, the contents of these chapters are explained in detail. Therefore, due to the limited number of pages, I had to limit my discussion of specific applications to simple and typical ones, which were described in Chap. 5. The applications to a wider range of more complex reactions will be discussed at another time. In Chapter 6, the basic features of new reaction kinetics are summarized.

This book is intended mainly for graduate students, young researchers, and teachers in science and engineering, but it can be understood by undergraduate students as well as liberal arts students if they read it in order from Chap. 1. Experts would be able to understand the contents of this book by starting from Chaps. 3–6 and referring to Chaps. 1 and 2 as necessary.

In this book, I have tried to discuss the reaction kinetics from a consistent standpoint. Unfortunately, in Japan, there are not enough opportunities to discuss one's standpoint on any subject sufficiently thoroughly to expose each position. The critical inheritance of reaction kinetics and its development described in this book is only an attempt, and as the author, I would be more than happy if the issues raised in this flawed book could trigger discussions in seminars, study groups, reading groups, etc., on reaction kinetics, the fundamentals of science in general, and issues of scientific cognition.

Ichijoji, Kyoto, Japan
December 1979

Satoru Yamamoto

Contents

About the Author and His Bibliography

Dr. Satoru Yamamoto was born in 1941. He received doctoral degree from Kyoto University in 1970 and continued his research for 35 years at the same institution. He has achieved notable results in variety of research fields such as Graphitization of Cementite (10), Spheroidization of Graphite (20), Martensitic Transformation (3), Aging Precipitation (6), Reaction Kinetics (5), Molecular Orbital Calculation and Alloy Theory (10), Microstructural Control of Alloys and Alloy Design (8), High Abrasive-resistant Spherical V-C Carbides (9), and Others (7). The number of published articles in scientific journal is shown in parentheses. Representative articles written in English associated with this book and books written by Dr. Yamamoto are introduced here (H. Yoshida).

Journal Articles

(1) S. Yamamoto, Y. Kawano, N. Hattori, Y. Murakami and R. Ozaki: Influence of hot impact deformation on graphitization in white cast iron, Metal Science, 11, 571–577 (1977). https://doi.org/10.1179/msc.1977.11.12.571
(2) K. Kubota and S. Yamamoto: Kinetics of Graphitization of Cementite, Trans. JIM. 27, 328-340 (1986). https://doi.org/10.2320/matertrans1960.27.328
(3) S. Yamamoto: The Time Energy Uncertainty Principle and Thermal Activation, Zeitschrift für Physikalische Chemie., 290, 17–32 (1989). https://doi.org/10.1515/zpch-1989-27003.
(4) Yamamoto: Cohesive Energy and Energy Fluctuation as a Measure of Stability of Alloy Phases, Acta Materialia, 45, 3825–3833 (1997). https://doi.org/10.1016/S1359-6454(97)00045-1

Books

(1) S. Yamamoto: **New Reaction Rate Theory — Beyond Absolute Rate Theory** — (Showado, Kyoto, 1979).
(2) S. Yamamoto and T. Tanabe: **Energy, Entropy and Temperature**, (Showado, Kyoto, 1981).
(3) **Spheroidal Graphite Cast Iron**, Edited by H. Cho, S. Akechi and K. Hanawa, (AGNE, 1983).

(4) S. Yamamoto and T. Tanabe: **Science and Structure of Cognition**, (Showado, Kyoto, 1984).

(5) S. Yamamoto and T. Tanabe: **New Approach to Materials Science — Unified understanding based on quantum mechanics —**, (Showado, Kyoto, 1990).

(6) N. Inoyama, S. Yamamoto and Y. Kawano: **Cast Iron based on Reaction Theory**, (Shin-nihon Casting and Forging Association, 1992).

(7) N. Inoyama, S. Yamamoto and Y. Kawano: **Cast Irons Clarified through Bonds and Reactions**, (Chinese Academy of Sciences, 2000).

(8) S. Yamamoto: **Democritus' Atomism and Modern Metallurgy — Electron-photon interaction and bonding, structure, properties, and reactions in materials —**, (Showado, Kyoto, 2005).

(9) Spheroidal Material Research Consortium: **Stainless Spheroidal Carbide Cast Material**, edited by S. Yamamoto and T. Tanabe, (Nikkan Kogyo Shimbun, Tokyo, 2006).

(10) M. Takeda, K. Hanawa, T. Tanabe and S. Yamamoto: **History and Logic in Alloy Theory — Using the data of "Handbook of Metals" and "Handbook of Iron and Steel"**, (Muse Corporation, Niigata, 2007).

(11) S. Yamamoto and K. Hanawa: **Struggling with Contradictions — Correction of Misunderstandings**, (Kenbunsha, Kyoto, 2015).

(12) S. Yamamoto and K. Hanawa: **Preparation against Contradictions**, (Kenbunsha, Kyoto, 2016).

(13) S. Yamamoto and K. Hanawa: **The Future of Materials Science — From Pairing of Valence Electrons and Unoccupied Valence Electrons —**, (Muse Corporation, Niigata, 2019).

Part I
Basis for Construction of Our Reaction Kinetics

Chapter 1
What is Reaction Kinetics as a Scientific Cognition?—Toward Innovating Conventional Reaction Kinetics

Abstract Reaction kinetics is a theory of changes over time in natural phenomena, and the theory begins with human understanding of the origins of the world (nature). Since the nature consists of various layers of hierarchy and changes over time, human understanding of nature must also be based on various concepts from these two aspects of hierarch and historicity. In this chapter, the characteristics of quantum mechanics, classical mechanics, and thermodynamics are first described, and the relationship between the three disciplines is derived from statistical averaging operations, either classical or quantum, with statistical physics as a mediator, and quantum mechanical averaging operations as the more fundamental. In addition, concepts such as non-equilibrium, non-stationarity, irreversibility, uncertainty, discontinuity, and so on are important to reaction kinetics and are discussed with each counterconcept. These conceptual considerations suggest the logic in this chapter that the time–energy uncertainty principle of quantum mechanics, which explicitly contains time, is necessary to theoretically construct a full-fledged theory of our reaction kinetics.

Keywords Reaction kinetics · Hierarchical-historical movement · Uncertainty principle · Quantum mechanics · Classical mechanics · Thermodynamics

1.1 Introduction

The purpose of this book is to clarify underlying problems in the current kinetics as a whole, to identify the way how reaction kinetics should proceed in the future, and to show a new attempt along this perspective. What is reaction kinetics? "The answer is in the question."—"Zen Buddhism" saying. It means that questions never appear suddenly without any relation to the answer and also answers are never found by decomposing the questions from outside of them. When you ask a question, the answer is already involved in the question to some extent. In fact, it must be said that how we ask questions and what we ask are inextricably linked to the kind of answers we are seeking. What questions will be addressed and how will they be answered are quite dependent on what to think about reaction kinetics, what and how to ask about reaction kinetics.

S. Yamamoto, *Reaction Kinetics Based on Time-Energy Uncertainty Principle*,
https://doi.org/10.1007/978-981-19-9673-3_1

In the present chapter, therefore, we would like to deeply ask from the broadest possible perspective what reaction kinetics is. Even if this way is a bit of a detour, the harvest will be rich and fruitful, and let us be willing to take a detour. You may be able to hope that the discussion in this chapter will be useful as a viewpoint for the critique of the absolute reaction kinetics in Chap. 2 and as a guideline for the construction of a new reaction kinetics in Chap. 3.

1.2 Hierarchical and Historical Development of World—Our Cognition

The world in which we mankind live is expanding with history and will continue to grow. Such a spread of the world is not just a spread but has two aspects: hierarchical (spatial) and historical (temporal) structures. At first, the world of mankind was limited to a human-level world (macro-world) that could be directly seen and touched by humans. With the development of history, however, mankind has come to know the existence of molecules and atoms and has actively exploited the phenomena related to these molecules and atoms. It is now known further that molecules and atoms are composed of more basic particles such as protons, neutrons, and electrons. It may be revealed in the future that even elementary particles such as electrons have an internal structure. Such hierarchical spread of the world extends not only to the direction of the "micro" and "submicro", but also to the "macro" and "supermacro" such as the earth, star clusters, galaxies, galaxy clusters, and supergalaxies.

The reason why the world of mankind (human being) expands in this hierarchical way is that the world (nature) itself has such a hierarchical structure. The world (nature) movement has a historical (temporal) aspect in addition to such a hierarchical (spatial) aspect. That is, modern science has revealed that since the creation of the universe, it has historically generated, developed, and evolved from elementary particles to atomic nuclei, atoms, molecules, inorganic substances, organic substances, plants, animals, and then mankind. In this way, the world (nature) that includes us mankind is in a hierarchical and historical movement. Since the world is undergoing a hierarchical and historical movement, the spread of our world has been and will continue to be hierarchical and historical.

Our world expands hierarchically and historically. But the expansion of our world does not mean that the world will expand without our action of anything about it. To expand the world of mankind means to expand the world in which mankind works and to make the world (nature) work for us. In order for mankind to expand the world of its activities and to make non-our world our world, we must know the laws of movement in each hierarchical level and in each historical process of the world. In other words, there exists the problem of cognizing the hierarchical and historical laws of movement of the world behind the expansion of our world. Then, how has mankind cognized the hierarchical and historical laws of movement in this world? What do we mean when we cognize things? In the following, we will consider the problem of cognition of the things.

When we try to cognize the world (nature), the specific content of that cognition is naturally defined by the object of cognition. Since the world (nature), the object of cognition, is in hierarchical and historical movement, its cognition must be hierarchical and historical. Because of the hierarchical nature of world movements, our cognition must necessarily be structural. Therefore, it is necessary to disassemble a composite or a structure into constituent elements, and conversely to construct and assemble a composite or a structure from those elements. The above process of thinking decomposes a substance into its constituent, i.e., atoms, and conversely examines the structure of how the atoms combine to form a substance. By repeating such structural pursuits, complex materials are broken down into simple ones, which are further broken down into simpler ones.

In this way, cognition of multiple hierarchical structures is obtained. And finally, if it is possible to reach elementary particles that cannot be further decomposed, all other substances will be constituted by such elementary particles. On the other hand, the historicity of the world argues that it is necessary to cognize the development of the world, or the motions, changes, and reactions of matter. Such a movement of substances is understood as a time variation of its state, the future is understood by the past and the result by the cause with linking a later state in time to an earlier state. The results of cognition are then summarized in a form of the law of motion and the law of causality. Thus, our cognition about the world increases its hierarchical and historical breadth and depth, due to the hierarchical and historical development of the world.

1.3 Scientific Cognition—Axioms of Objectivity and Criteria of Truthfulness

In the previous section, the readers would recognize such viewpoints that the world (nature), which is the object of cognition, is performing a hierarchical and historical movement, and therefore, that the cognition must be a hierarchical and historical. However, it is not always allowed to use any kind of cognition as long as cognition of the world is hierarchical and historical. Actually, we are aiming for a scientific cognition of the hierarchical and historical movements of the world. Comparing other fields of science, such as religion, philosophy, and ethics, to the realm of science, what makes science the very "science" is that it is based on the "axioms of objectivity" and "criteria of truthfulness" that science presupposes.

It is well known that science puts the logical consistency and the possibility of experimental verification as a criterion of its truthfulness. However, this criterion of truthfulness alone did not make the cognition scientific. This is evident in the historical context where modern science could not be established by pre-Galilei or pre-Descartes way of cognition. Certainly, it is not that pre-Galilei and pre-Descartes scholar ignored reason, logic, or experimentation, or they did not think of organizing and matching them. However, "science" in the sense we understand today could not be constructed based on these alone. Galilei and Descartes denied the conventional teleology that interpreting phenomena from the viewpoint of their intrinsic purpose allows us to arrive at a cognition of truth. According to their postulate, there is not a "purpose" or a god in the universe, and nature is an objective being. By accepting this postulate of objectivity and abolishing Aristotle's "Physics" (natural philosophy) and cosmology, the foundation for cognizing modern science has been laid.

1.4 Progress and Innovation in Scientific Cognition

Looking at the innovation in scientific cognition about the hierarchical and historical movement of nature, we can recognize some characteristics. The first is a change in the method of scientific logic from induction to deduction. In each field of science, at their early days, the method of induction is often relied upon in combination with the scientific idea of respect for facts. In contrast, the use of deductive methods was relatively rare. A typical example of the deductive theoretical system is Lagrange's analytical mechanics. His theory of mechanics was an example of a theoretical system based on the belief that simply developing formulas would give all the equations needed to solve the problem.

The wealth of observations accumulated by inductive methods was then organized into a logical system of scientific theory based on a few well-established laws of motion. In electromagnetics, such a change has been completed from Faraday's inductive theory to Maxwell's deductive one. Such a revolution has started in the field of chemistry since the emergence of quantum mechanics. Quantum mechanics was founded between 1925 and 1926, and its effects have begun to emerge gradually. Textbooks of chemistry have been completely rewritten from inductive way to deductive one after this gradual change.

The second characteristic of this progress in scientific cognition is the shift from qualitative to quantitative description. Scientific cognition always begins with the discovery of new phenomena or new qualities. First, a qualitative observation is made. However, rather than staying at a level of a qualitative description of phenomena, quantitative measurements are gradually being made, and eventually, leading to the form of quantitative laws. In this way, when scientific cognition progresses from inductive way to deductive one and from qualitative way to quantitative one, the content of that cognition comes to be expressed by mathematical equations of motion. At this point, scientific cognition has dramatically increased its power from the passive interpretation of nature to the positive prophecy and reaches a stage where it can be incredibly powerful.

In addition to the progress of scientific cognition mentioned thus far, we must point out that our historical experience is not only mere progress of cognition, but also what can be called cognitive innovation. Einstein's theory of relativity is the first revolution in scientific cognition brought about by the various discoveries in the twentieth century. The revolution in scientific cognition is caused by not only the theory of relativity but also quantum mechanics. Here, let us consider not the contents of the theory of relativity or quantum mechanics, but the meaning of their appearance affected on the scientific cognition.

For example, the idea of relativity was at first strongly opposed by physicists and philosophers. Physicists have approached this new hypothesis with a critical spirit. In order for the new idea to be accepted in physics, it had to pass the following three-step verification. First, the new idea must not disturb what has been successful in the past work, but must inherit the past idea as a result, and must not overturns the explanation of the observations that have been used to support the theories so far. Second, it must explain the new evidence in a convincing way that makes old ideas questionable and suggests new ones. Third, it must anticipate new phenomena or new relationships that were not known or were not clearly understood at the time when it was conceived.

The forms of scientific progress and innovation mentioned above can be considered as follows. The theoretical system of science L is accompanied by a range of experience A to which it applies. If we apply this L to a wider range of experience than A, we may find a discrepancy between experience and theory in the application. And a new theory L' is found that eliminates this discrepancy. This new theory L' is also accompanied by its range in application of experience A'. A' contains A. Such process is the evolution of theory that the transition from L to L' is repeated one after another [1, p. 5].

1.5 Reconstruction of World in Terms of Our Concepts

Cognition of things has two aspects: the object of cognition and the subject of cognition. Until the previous section, the problem of cognition was mainly considered from the side of the object to be cognized. In this section, let us consider the problem of cognition from the human side as the subject of cognition.

Cognition can be defined as the conceptual reconstruction of hierarchical and historical movements of the world, the object of cognition, by humans, the subject of cognition. Therefore, the hierarchy and history of the nature as the object is reflected in the conceptual cognition system of humans and has its counterpart. The hierarchical aspect of the nature is reflected in the conceptual system of mankind as follows.

The nature has a hierarchical structure. That is, there is such a multi-layered hierarchical structure in the nature that a complex substance is composed of simple substances, while this simpler substance is composed of further simpler substances and so on. By repeating this hierarchical cognition, we finally reach elementary particles. Reflecting this hierarchical structure of the outside world, there is also a hierarchical structure in our conceptual system. Some concepts are complex, and others are simple. Complex concepts are described by simple concepts, and simple concepts are described by simpler concepts, and so on. There exists “elementary concept” for materialistic elementary particles. And all other complex concepts are explained by this "elementary concept". The elementary concept itself is assumed to be self-evident without definition and explanation.

On the other hand, the nature has historicity. That is, the nature changes over time, as it evolves, moves, reacts, and so on. During the change in the state of matter, the later state in time arises based on the preceding state in time. The future is brought about by the past, the consequences by causes, and thus the changes are temporal and causal. The counterpart of such historical change of the nature is found in the logical form of thinking of mankind. This is the "principle of sufficient reason" in inference. We derive our consequences in logic from its reasons. Our theoretical system is a hierarchical and historical reflection of nature, including its counterpart. This system reflects the outside world, but relatively independent of the outside world. The concepts and logics are the intellectual property resulting from the historical cognitive activities of mankind, which will be continuously refined by cognitive activities and become richer and richer in meaning.

In the following, let us consider in more detail the characteristics of the reflection and response of this hierarchical-historical movement of the external world to the conceptual logical system of the internal world. Our cognitive actions will start from perceptive understanding by reducing the unknown to the known, the complex to the simple, and the unfamiliar to the familiar. However, what are known, complex, and familiar things to understand unknown, complex, and unfamiliar things? In a word, we can dare to say that they are something that can be obtained directly from our sense as mankind, or something we are accustomed to in our daily lives. In other words, these things that should be the basis of our cognition are defined by the fact that we, the subject of cognition, are mankind, and therefore defined by our sense.

However, the basis of our cognition is not absolute and the criterion of which may be changeable. What is simple and straightforward may change historically. For the time being, we will not go any further into what is simple and understandable to us. Let us be satisfied that we understand everything based on macroscopic physical quantities obtained by our senses and observation instruments. Thus, when we try to understand the hierarchical and historical aspects of the world, we always understand it in terms of our human-level words, *or* in terms of macroscopic concepts. This means that even the motion in the micro-level of atoms or electrons must be described with macro-concepts. The fact that the current recognition system is actually doing so is evident from the circumstances described below. For example, the concepts and physical quantities adopted in quantum mechanics are ultimately classical and macroscopic. This is because all measurement tools for studying microscopic objects are macroscopic and nothing can be done without it. Since humans are macroscopic beings in the first place, all the parameters related with us are ultimately attributed to classical and macroscopic ones.

We, human beings, cognize the hierarchical and historical movements of the world, as stated above. Therefore, the cognition is performed both hierarchically and historically, based on the humans-living hierarchy and on the human history, and thus focused on what is easy for human beings to understand. This is an extremely important premise for cognition. This fundamental premise has crucial implications for the cognition by human beings.

The **first** is the constraints on what we cognize. We are forced to start with what is familiar to us and easy to understand. For example, we prefer to study motionless over motion, equilibrium over non-equilibrium, reversible over irreversible, steady over unsteady, regular over irregular, and certainty over uncertainty. To put this in other words from a hierarchical and historical perspective, this can be said that our cognition will start from any things or events close to the human hierarchy and history. Too micro-hierarchy for humans such as elementary particles and too macro-hierarchy such as galaxy clusters and superclusters are unknown areas to us since they are far from our human hierarchy. Similarly, we are still largely unaware of events that are too far away for mankind, i.e., events of extremely short or extremely long durations. These circumstances indicate which way we must struggle if we are to expand our cognition.

The **second** is the constraint on elevation of our cognition into a conceptual and logical system. We expand our cognition from "world" that is hierarchically and historically closer to the realm of mankind and is easier for humans to understand to "world" which is farther away from the realm of mankind and is incomprehensible to us. The content of newly added cognition is elevated into a conceptual and logical system. What is important when micro-hierarchies such as atoms and electrons are the object of cognition is to make a clear distinction between microscopic and macroscopic concepts and physical quantities, and to clarify the relationship between them. Thus, there are various systems in our cognition depending on the hierarchical and historical aspects of the world (nature).

1.6 Science and Mathematics

In the previous section, we have shown the "principle of sufficient reason" in inference which means logical forms of thinking for temporal and causal changes in the external world. The most typical example is mathematics. Mathematics is a pure form of thinking obtained by abstracting our way of thinking. Kant used the term "a priori" to define knowledge that is independent of experience. However, "knowledge independent of experience" does not mean knowledge which does not rely on any experience in a genetic, historical, or psychological sense. He knew well that all human knowledge relied on experience in a genetic sense. Obviously, there exists no knowledge of any kind without experience. Nevertheless, Kant used the term "a priori" as opposed to the term "a posteriori" to show that mathematics is the system of knowledge in which we owe the least to the external world.

Mathematical objects, unlike those of physics, have no direct counterpart in experience. The correspondence between mathematics and experience is an indirect and genetic correspondence. Therefore, the criterion of truthfulness in mathematics is different from that of empirical science (especially physics) mentioned earlier and is based on consistency and generality of formal laws. This criterion does not include the experimental verifiability to guarantee a direct correspondence with experience.

Therefore, mathematics has the following characteristics:

(1) What thinking must inevitably follow, and a priori controlling principle of thought,
(2) Something that is not a direct result of science, but makes the experience, the premise of science, possible, and
(3) A principle that is testable in experience, but is not derivable from experience, nor can be further analyzed.

Thus, a special relationship between empirical science and mathematics emerged, through which the progress of science encouraged the invention of new mathematical tools, while the progress of mathematics provided rich forms of mathematical methods and various contents are related to the forms. Science uses various mathematical forms (differential equations, integral differential equations, finite difference equations, matrix equations, arithmetic laws, algebraic laws, integration principles, etc.). Mathematical formulation makes events scientific. Conversely, unless the phenomena studied are expressed in the form of mathematical laws, they cannot declare the problem has been solved. We should make great use of sharpened form of mathematical thinking.

Nature can be described with help of mathematical concepts. Mathematical symbols make the physical concepts clear and unambiguous. However, mathematical symbols are not a substitute for physical concepts. Physics cannot be reduced to mathematics, because it is evident from the differences in the criteria of truthfulness between empirical science and mathematics. In other words, mathematics is formal and just a sign game. It is tautological, uninformative, and not directly related to the external world. It does not reveal anything about the external world. It simply gives coordinates to the physical laws obtained by experiments.

Considering the nature of mathematics mentioned above, when expressing cognition of the outside world using mathematics, the following is necessary. That is, we must clearly define the correspondence rules between physical concepts and mathematical symbols, or physical laws and their mathematical expressions. Such a situation means that our scientific cognition and its mathematical expression are related as follows.

$$\begin{aligned}&(\text{Physical Concept, Laws}) \leftarrow (\text{Correspondence Rule})\\&\quad \rightarrow (\text{Mathematical Formal Expression})\end{aligned}$$

And the theoretical system of scientific cognition is as follows.

$$\begin{aligned}&(\text{Physical Law}) + (\text{Logical Law})\\&\quad = (\text{Scientific Theory: Mathematical Formal Logic System}).\end{aligned}$$

1.7 Hierarchy and Conservation Laws

Up to the previous section, following problems have been described:

(1) the problem about the hierarchical and historical movement of the external world, which is the object of cognition, and
(2) the problem about the construction of a scientific theory as a sophisticated logical system with mathematical forms through the cognitive actions by the subject, mankind.

As is clear, such a system of cognition is formed as a theoretical system that is unique to the hierarchy to which the object of cognition belongs. Figure 1.1 presents a brief overview of such a situation in the field of physics that we are interested in for the time being [2, p. 295].

As described above, the unique cognition system corresponds to hierarchy of physical motion of the object. Today, many laws of conservation are found in various fields of physics. They are categorized as seen in Fig. 1.2.

In this classification, the general law of conservation is a physical law of conservation that is valid for all levels of hierarchies of matter currently recognized, and this law is the basis for the unification of physical knowledge. In contrast, the partial laws of conservation are valid only in a limited hierarchy and represent the peculiarity of individual hierarchies. The laws in the new hierarchies and new aspects of nature, that may not be fully understood even in the future, involve the new conservation laws, and these new conservation quantities are the expression of the nodes that link the new hierarchies and new aspects, distinguishing them from other hierarchies and

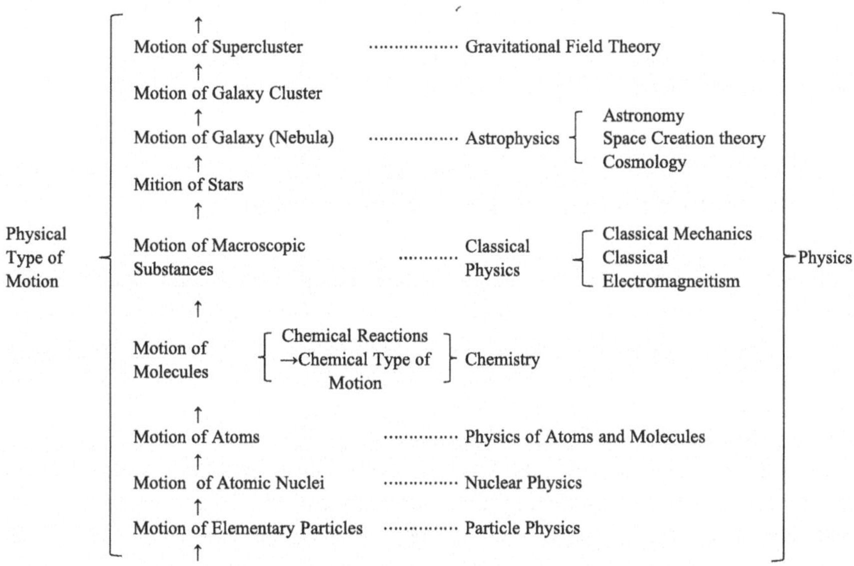

Fig. 1.1 Hierarchical motion of the world and scientific theories unique to the hierarchy [2]

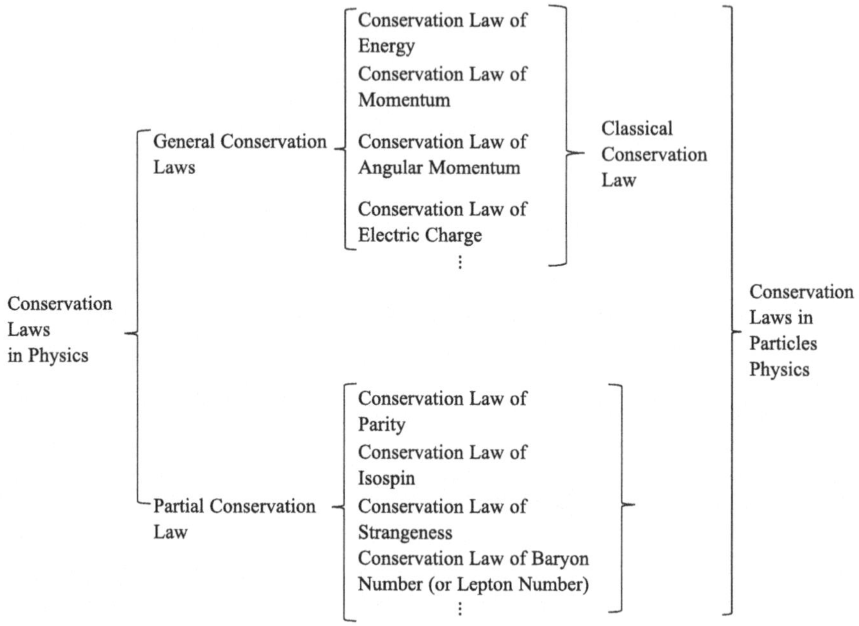

Fig. 1.2 Categories of conservation laws in various fields of physics

aspects. As our cognition expands across a wider hierarchy and wider dimensions, the nature of the conservation laws will change. The general conservation law that is now holds may be broken at some hierarchy or aspect in nature. However, it does not mean that the general law of conservation is lost but will be replaced by some other and more general law of conservation. Any concrete conservation laws are not absolute, but the idea of conservation itself would be absolute.

1.8 Reaction Kinetics and Description Systems

In the previous sections, we have considered the issue of scientific cognition in general as well as reaction kinetics. In the following sections, we will go into more in-depth considerations, limiting the range to those related to reaction rates. According to Linus Carl Pauling [3, p. 6], chemical bonds are said to exist when a force acts between two atoms or two groups of atoms, resulting in the aggregate formation, and the stability of the aggregates is large enough for chemists to treat this aggregate as a single independent molecular species. According to this definition, chemical reaction kinetics is a theoretical system that deals with the rate at which the state of the chemical bond changes from an old bonding state to a new one. It seems to involve three following problems in the chemical reaction kinetics.

First, since the hierarchy involved in chemical reaction kinetics is the atomic and electronic hierarchy, chemical reaction kinetics is a matter how we humans cognize the hierarchy of atoms and electrons. The body of theoretical system should naturally include the concept of the hierarchy of atoms and electrons and the concept of the hierarchy of human beings. It should be a knowledge system that connects these two concepts.

Second, chemical reactions are irreversible changes toward one direction. Therefore, as far as the chemical reaction kinetics is dealt, its basis must include something that can explain the irreversibility.

Third, rather than chemical reaction kinetics, the system must clearly describe the changes in chemical bonds with time.

Currently, the disciplines specifically related to reaction kinetics are quantum mechanics, classical mechanics, thermodynamics, and statistical physics. Quantum mechanics is a system that mainly describes the motion of atoms and electrons in microscopic hierarchy. Classical mechanics and thermodynamics are systems that describe our human world, macro-hierarchy. Statistical physics is a system that connects these two. In the following, we will analyze the contents and logical characteristics of these four theoretical fields of physics. At this time, three issues mentioned above related to chemical reaction kinetics are discussed;

(1) the hierarchy of matters to be treated,
(2) irreversibility, and
(3) the explicit expression of the time rate of change.

1.8.1 Logic of Quantum Mechanics

Quantum mechanics is a theory that describes the dynamical phenomena and time-dependent changes of states of individual particles with the quantities (momentum, energy, angular momentum, etc.) almost the same as the Planck constant h ($= 6.62607015 \times 10^{-27}$ erg s) and consists of two parts: the wave equation and observations.

Wave Equation

1. Let us consider the characteristics of the wave equation. According to de Broglie, all matters have both properties of particle and wave, and the relationship of $\lambda = h/p$ is established between the momentum p as a particle of a matter and the wavelength λ as a wave. If the potential energy of the field where a particle is moving changes remarkably within a range smaller than the magnitude of this wavelength λ, the behavior of this particle becomes wavelike, and its state is represented by a wave function Ψ, and the change of the state with time is expressed by

$$i\hbar\frac{\partial\psi}{\partial t} = \hat{H}\psi \tag{1.1}$$

where i is the imaginary unit $i \equiv \sqrt{-1}$, $\hbar \equiv \frac{h}{2\pi}$, and $\hat{H} \equiv -\frac{\hbar^2}{2m}\left(\frac{\partial^2}{\partial x^2} + \frac{\partial^2}{\partial y^2} + \frac{\partial^2}{\partial z^2}\right)$ is an operator called Hamiltonian. That corresponds to the total energy of classical mechanics:

$$\frac{1}{2m}\left(p_x^2 + p_y^2 + p_z^2\right) + V$$

and can be obtained using substitution of (p_x, p_y, p_z) by $(\frac{\hbar}{i}\frac{\partial}{\partial x}, \frac{\hbar}{i}\frac{\partial}{\partial y}, \frac{\hbar}{i}\frac{\partial}{\partial z})$. Ψ is a wave function and t is time.

The followings point out some characteristics of this wave equation.

The first is the symmetry of this equation with respect to time reversal, or in other words, the reversibility of the motion this equation describes. In general, a wave function Ψ can be represented such that its difference from its complex conjugate is only the sign of time. Therefore, if t is changed to $-t$ in Eq. (1.1) and Ψ is replaced by its conjugate complex function, Ψ^*, the wave equation preserves its forms. This feature of the equation means that if the state changes in a certain order, the state can change in a completely opposite order.

The second is the conservation of energy in the stationary state. If the system is not placed in a time-dependent external field, the Hamiltonian does not explicitly include time. This is immediately derived from the fact that all times are equivalent for a given physical system if no external field exists (*or,* under a static external field). In systems that are not placed in a time-dependent external field, the Hamiltonian function is conserved, which is called "energy". This is the law of conservation of energy in quantum mechanics, which means that if the energy takes a definite value in a certain state, that value is constant over time. The state where the energy takes a definite value is called the "stationary state" of the system. In contrast, if the Hamiltonian is explicitly time-dependent, energy is generally not conserved, and there is no stationary state. In the stationary state, we can assume Ψ would be

$$\Psi = \varphi \exp[-iEt/\hbar] \tag{1.2}$$

where E is a constant energy, and φ is a function of the coordinates of the configuration space but not including time. By substituting Eq. (1.2) into Eq. (1.1), we can obtain

$$\hat{H}\varphi = E\varphi \tag{1.3}$$

This is called the time-independent Schrödinger equation.

If a system consists of two or several parts, and there is no interaction between those subsystems, the followings can be easily shown. The Hamiltonian of the whole system is the sum of the Hamiltonians of subsystems as

$$\hat{H} = \hat{H}_1 + \hat{H}_2 + \cdots + \hat{H}_n \tag{1.4}$$

where $\hat{H}_1$ contains only coordinate q_1, $\hat{H}_2$ contains only coordinate q_2, and $\hat{H}_n$ contains only coordinate q_n. In this case, we can express the eigenfunction φ of the operator $\hat{H}$ by the product of the eigenfunctions $\varphi_1, \varphi_2, \ldots, \varphi_n$ of $\hat{H}_1, \hat{H}_2, \ldots, \hat{H}_n$ like

$$\varphi = \varphi_1 \cdot \varphi_2 \cdot \ldots \cdot \varphi_n \tag{1.5}$$

And the energy eigenvalue is the sum of the eigenvalues of each operator as

$$E = E_1 + E_2 + \cdots + E_n \tag{1.6}$$

Here, it should be noted that the physical content that there is no interaction between subsystems can be mathematically expressed as a sum or a product.

2. Let us describe specific applications of these wave equations. Applications to the study of physical systems can be broadly divided into two cases where the Hamiltonian is time-dependent and time-independent. When the Hamiltonian is time-dependent, there is no stationary solution of the Schrödinger equation (1.1). As a method often used in this case, there is a time-dependent perturbation theory, which is sometimes called "method of variation of constants". In this case, it is assumed that the Hamiltonian $\hat{H}$ is divided into two parts as $\hat{H} = \hat{H}_0 + \hat{H}'$, where $\hat{H}_0$ is a part that does not include time, and the eigenvalue equation $\hat{H}_0 \Psi = E_0 \Psi$ can be solved exactly, while $\hat{H}'$, which is time-dependent, is assumed to be a small external disturbance to this system. As another method, the adiabatic approximation is used for large but slowly changing disturbances, and sudden approximation is used when the Hamiltonian suddenly changes from one form to another. These specific applications are left to Chap. 3.

Most applications of quantum mechanics to physical systems are for the case where the Hamiltonian is time-independent, and there are two main types of problems:

(1) Determination of the energy level of the bound state when E in Eq. (1.3) is negative. This is to find eigenvalues belonging to the discrete spectrum of Hamiltonian.

(2) Determination of collision cross section. This results in finding the asymptotic form of the eigenfunction for the unbound state when E in Eq. (1.3) is positive.

There are not so many systems that are physically interesting, and wave equations can be exactly solved. In practice, approximations often play an important role, and there are various methods of approximation depending on whether the problem deals with bound or unbound (scattered) states. Typical examples include stationary perturbation theory, variational methods, series treatment of perturbations, and Wentzel–Kramers–Brillouin approximation (abbreviated as WKB approximation).

(3) **On the Basis of Quantum Mechanics**, there is an uncertainty relation that is established between physical quantities that are conjugate to each other, *for*

example, position x and momentum p. We can usually define the uncertainties in x, Δx and p, Δp by their standard deviations [4, p. 133], as

$$\Delta x \equiv \sqrt{\langle (x - \langle x \rangle)^2 \rangle},\ \Delta p \equiv \sqrt{\langle (p - \langle p \rangle)^2 \rangle} \tag{1.7}$$

where parentheses, $\langle\ \ \rangle$ represent the mean values. Then we consider

$$I(\lambda) = \int_{-\infty}^{\infty} \left| x\varphi + \lambda\hbar \frac{\partial \varphi}{\partial x} \right|^2 \mathrm{d}x \tag{1.8}$$

which is always positive or zero for any value of λ. By expanding and partial integration, we can obtain

$$\begin{aligned} I(\lambda) &= \int_{-\infty}^{\infty} |x\varphi|^2 \mathrm{d}x + \lambda\hbar \int_{-\infty}^{\infty} \left(\frac{\partial \varphi^*}{\partial x} \cdot x\varphi + x\varphi^* \frac{\partial \varphi}{\partial x} \right) \mathrm{d}x + (\lambda\hbar)^2 \int_{-\infty}^{\infty} \left| \frac{\partial \varphi}{\partial x} \right|^2 \mathrm{d}x \\ &= \int_{-\infty}^{\infty} \varphi^* x^2 \varphi \mathrm{d}x - \lambda\hbar \int_{-\infty}^{\infty} \varphi^* \varphi \mathrm{d}x - (\lambda\hbar)^2 \int_{-\infty}^{\infty} \varphi^* \frac{\partial^2 \varphi}{\partial x^2} \mathrm{d}x \end{aligned} \tag{1.9}$$

For simplicity, we can assume $\langle x \rangle = 0$ and $\langle p \rangle = 0$ in Eq. (1.7). Then,

$$I(\lambda) = (\Delta x)^2 - \lambda\hbar + \lambda^2 (\Delta p)^2 \geq 0. \tag{1.10}$$

The condition that this quadratic expression of λ is not negative is that the discriminant expression is not positive, which leads to

$$\Delta x \cdot \Delta p \geq \frac{\hbar}{2}. \tag{1.11}$$

This relation is called "Heisenberg's uncertainty principle", which means that physical quantities that are conjugated to each other cannot have a definite value at the same time. The establishment of such an uncertainty relation is one of the most distinctive features of quantum mechanics, and no matter how strange this property may look, it is clear from the above proofs after all. As described above, it is derived from the particle and wave duality of the matter.

Observation, Measurement Problem

Now that we have described one of the two components of the quantum mechanical theory system, that is, the matter relating to the wave equation, let us talk about another component, the measurement problem. In quantum mechanics, when a state represented by a wave function is observed, the relationship between the observed values and eigenvalues of the corresponding operator is formulated:

(1) Measurement of one quantity (coordinate, energy, etc.) generally gives a unique value that appears as an eigenvalue of the corresponding operator.
(2) If φ is expanded by using a perfect orthogonal system of eigenfunctions of physical quantity G as

$$\varphi = a_1\psi_1 + a_2\psi_2 + \cdots \tag{1.12}$$

and

$$G\psi_K = \lambda_K\psi_K \tag{1.13}$$

then, the probability that the measurement gives value of λ_K is expressed by $|a_K|^2 = |(\psi_K, \varphi)|^2$. Quantum mechanics gives a fixed answer like classical theory only when the system is in a stationary state (isolated state) and $G\varphi = \lambda_K\varphi$. The probability for λ_K in this case is 1 and the measurement in this case gives λ_K with certainty.

(3) When a certain quantity is measured and found to have a certain value, the same value is obtained only if this measurement is repeated fast enough there. The wave function after the measurement, which actually gives the eigenvalue λ_K to G, must be the eigenfunction of G belonging to λ_K. The repeated measurement of G then gives again λ_K.
(4) The wave function changes during the measurement. When measuring G, the wave function shifts to one of the eigenfunctions of G. When the measurement gives the result λ_K, the wave function shifts to ψ_K. It is generally not possible to reliably predict to which eigenfunction of G the state function φ of the system will shift. In quantum mechanics, it only gives the probability of the transition of the wave function φ to ψ_K under the measurement of G, which is calculated from φ and ψ_K by $\left|(\psi_K, \varphi)^2\right|$. And importantly, this observation is essentially irreversible.
(5) The wave function changes in two very different ways, as stated below.
   (i) Continuous and causal reversible changes according to $i\hbar\frac{\partial\varphi}{\partial t} = H\varphi$
   (ii) Discontinuous/non-causal irreversible changes due to observation.

As described above, in order to describe the change of the wave function with time, quantum mechanics consists of two parts, one related to the wave equation and the other related to the observation. In that sense, there are still unsatisfactory points. Therefore, various discussions are still being made on this point, especially on the measurement problem. However, such deficiencies in the quantum mechanical system are not peculiar to quantum mechanics, nor do they imply that any such imperfection will reduce the value of quantum mechanics.

Now, a brief summary of the nature of quantum mechanics based on the contents described above is as follows. **First** of all, in quantum mechanics the uncertainty principle holds, so there is quantum mechanical uncertainty. **Second**, as can be seen from the quantization of the eigenvalues of the Hamiltonian in the bound state, discontinuous changes in physical quantities occur. **Third,** as is clear from the relationship between the description using wave function and observations, the theoretical system is essentially stochastic in nature. In other words, this theoretical system is not a completely deterministic and causal system like classical mechanics, but a stochastic system.

1.8.2 Logic of Classical Mechanics

Classical mechanics refers to the theoretical system to describe dynamical phenomena and the change of the state of a particle with time in which the magnitude of the dynamic quantity (momentum, energy, angular momentum, etc.) of one particle is much larger than the Planck constant, h. Classical mechanics is a deductive system based on the following three basic laws.

< **First law** > (Galilei's law of inertia)—When an object is left alone, it moves in the same direction at a uniform speed.

< **Second law** >—Force (F) is equal to mass (m) multiplied by acceleration (α) as

$$F = m\alpha = \frac{\mathrm{d}p}{\mathrm{d}t} \tag{1.14}$$

where p means momentum and t is time.

< **Third law** >—Action and reaction are equal to each other.

The system of classical mechanics based on these three laws has the following characteristics of symmetry in space and time.

1. The classical mechanics equation is invariant with respect to time reversal, *that is*, a change in the sign of time. This symmetry in the two directions of time is manifested in the fact that changing the sign of time does not change the form of the equation. In this way, the motion described by the classical mechanics is essentially reversible, *that is*, if a change occurs in a certain direction, a change in the opposite direction is also possible. Furthermore, it should be pointed out that the conservation law that holds in classical mechanics is related to the symmetry of time and space as follows [5, Chap. 2]:

   Homogeneity in time (translational operation for time)

   $\Leftrightarrow$ the law of conservation of energy

   Homogeneity in space (translational operation for space)

   $\Leftrightarrow$ the law of conservation of momentum

   Isotropy of space (rotational operation for space)

   $\Leftrightarrow$ the law of conservation of angular momentum

2. Let us consider the characteristics of classical mechanics.

The **first** characteristic of this theoretical system is the definiteness of its physical quantity. In classical mechanics, unlike quantum mechanics, there is no uncertainty relation, and therefore a complete definiteness of physical quantities is assumed. Thus, the reason that quantum mechanical uncertainty is not a problem in classical mechanics is that the mechanical quantity of the target particle is much larger than Planck's constant h. In that case, the duality of matter plays no important role. Or conversely, classical mechanics deals with the case where the magnitude of Planck's constant h can be ignored, that is, only the phenomenon that h can be regarded as zero. In such a case, the wave nature of matter becomes insignificant and can be treated as so-called classical particles.

The **second** feature is that the physical quantities dealt with by this dynamic system are considered to change continuously. In other words, it is thought that there is no discrete value or discontinuous change such as the eigenvalue of Hamiltonian in the bound state of quantum mechanics. This also stems from the fact that the classical mechanics deals with only dynamic quantities whose magnitude is such that Planck's constant h can be regarded as zero.

Third, classical mechanics is a completely deterministic theoretical system in the sense that it can uniquely determine the state of the future when given an initial value and is in sharp contrast to the stochastic theoretical system of quantum mechanics.

Determinism of Classical Mechanics [6, Chap. 30].

Now, let us take a slightly more in-depth discussion of **this third feature**, that is, classical mechanics is completely deterministic. Since having learned classical mechanics, we have become accustomed to the idea that in this dynamic system, if given an initial value of the system, we can fully predict the subsequent dynamic system, and this conviction looks like a self-evident truth. However, there is actually involved an apparent false in the idea and, speaking more precisely, we cannot describe anything definitely but there should be any part of silence, that is an implicit premise.

First, the initial values are not determined indefinitely in practice. Strictly we only get a certain distribution of initial values, the value of which is known to us only at a certain probability ratio. In addition, the uncertainty of the initial value increases with time in the course of the motion, and the uncertainty of the future state will always exceed a conceivable size.

Second, during the motion of the system, our unforeseeable accidental forces will act on the system. Even if the accidental force is weak, the effect could be quite noticeable if long time passes. Therefore, the effects of such accidental forces cannot be ignored from real course of the motion.

Third, the question is whether the system is guaranteed to remain isolated over the future time that we are trying to predict. The prediction of classical mechanics presupposes that no force acts on the system during the time of interest to us, other than what the equation of motion of the system operates. Therefore, future dynamic systems can be predicted only if the isolation of the system is guaranteed in advance. Considering these three situations mentioned above, the theoretical system of classical mechanics is also an abstraction from reality. It can be said that it is effective only when making approximated predictions during not so long period for a system where isolation is guaranteed.

1.8.3 Logic of Thermodynamics

Thermodynamics is a logical system that deals with the equilibrium state of temperature-related phenomena in a system consisting of a huge number of particles and formulates the necessary conditions. Thermodynamics is a logical system formulated on the following three experience laws:

First law (the law of conservation of internal energy)—In an isolated system, matter and internal energy are preserved.

Second law (the law of increase of entropy)—Every process that occurs spontaneously in an isolated system goes in the direction of increasing the total entropy of the system.

Third law—the entropy of all objects in a stable or metastable equilibrium state decreases down to zero as the temperature approaches 0 K.

Isolated System

We have already used the term isolated system when considering about the logic of quantum mechanics and classical mechanics. Since the basic laws of thermodynamics hold for the isolated system, let us talk about the importance of the concept of this isolated system. An isolated system is, by definition, a system in which there is no exchange of energy or matter with the surroundings (the outside world). This concept of an isolated system is a basis for dealing with a non-isolated system, *that is*, a closed system or an open system. In other words, if the system under consideration is interacting with other systems to exchange heat or matter, the whole area of the system under consideration and the other system can be considered as one system again. This system is an isolated system, and the law that holds for an isolated system can be applied to this system.

Now, the concept of the isolated system requires a very important premise in the following points:

(1) Premise of applying the first and second laws of thermodynamics.
The entire thermodynamic theory is ultimately only valid for isolated systems.
(2) Premise of causality
As mentioned above, classical mechanics is a typical theoretical system that holds causality. In classical mechanics, solving the equation of motion and giving the initial conditions and boundary conditions give the state parameters of the system completely, and the future transition is uniquely determined as

long as the system is isolated. It is a deterministic and causal system. However, such circumstances are not limited to classical mechanics. Even in quantum mechanics, which is a stochastic theoretical system, the eigenvalue equations can be applied to isolated systems.

(3) There are seven conserved quantities for isolated systems.
The seven quantities are energy, momentum (3 components), and angular momentum (3 components).

(4) Source of additivity of physical quantity.
This point was mentioned in the section of "logic of quantum mechanics". If the system is divided into two subsystems that do not interact with each other, the Hamiltonian of the whole system is expressed by the sum of the Hamiltonians of the subsystems, so the energy of the whole system is equal to the sum of the energies of the subsystems. The same holds for thermodynamics. The additivity of energy and entropy is guaranteed by the system being isolated.
Furthermore, for isolated systems,

(5) It is a premise of statistical independence and universality of distribution functions in statistical theory. Thus, the idea of the isolated system is very important and the major premise of our theoretical system.

Equilibrium State

Next, let us talk a little about the concept of "equilibrium state", which is another important concept in thermodynamic theory. A macroscopic closed system under consideration is said to be in an equilibrium state in case its macroscopic physical quantities of any subsystem, which is itself macroscopic, are equal to their mean values with a great certainty. When the system is in the equilibrium state, there is no transfer or flow of state quantities inside the system. This concept of the equilibrium state is important in several respects:

(1) Prerequisites for the establishment of thermodynamic concepts, such as temperature and entropy, are concepts that can be defined only in the equilibrium state. Then, if the system is in an equilibrium state, the single-valuedness of the macroscopic physical quantity representing the state of the system is secured.

(2) If the system is in equilibrium, the principle of detailed balance can be applied. This principle can be generally expressed as follows. Even though a process may be complex in the sense that various independent mechanisms operate, the ultimate equilibrium when these mechanisms work together in a real system is the same in case any one of these mechanisms is uniquely involved. For example, a set of objects that were initially at different temperatures in an isolated enclosed area will reach the same final temperature, regardless of whether heat transfer between the different elements is attained by conduction or radiation. Furthermore,

(3) The equilibrium state is assumed for most laws related to thermal phenomena, such as the equation of state, the law of mass action, the Helmholtz equation, the law of equipartition of energy, the distribution law (Gibbs distribution for canonical ensemble or Boltzmann distribution), the Nernst theorem, and the principle of maximum entropy.

Characteristics of Thermodynamics

Then, let us consider the characteristics of thermodynamics.

1. **Science based on the macroscopic viewpoint.**

   What does the word "macroscopic" used here mean? "Macroscopic" is usually employed in such a loose way that we can see a matter directly with our eyes and touch it with our hands. In this section, therefore, we would take a deeper look at an essence of the concept of being macroscopic as an academic one. A macroscopic object has several characteristics:

   (1) **Very large mass**

   In the uncertainty relation $\Delta x \cdot \Delta p \cong \hbar$ in quantum mechanics (Δx and Δp are width of uncertainties in the position and momentum, defined by their standard deviations as in Eq. (1.7.), $\hbar$ is Planck's constant h (= 6.62607015 $\times$ 10^{-27} erg s) divided by 2π), $\Delta p = m\Delta v$. If the mass m is large, the widths of uncertainties Δx and Δv can be ignored. Thus, we can ignore the quantum mechanical uncertainties of various physical quantities of macroscopic object and can describe them in classical mechanics as having definite values.

   (2) **Huge number of particles and huge number of freedoms**

   For example, the temperature of one molecule is meaningless. The concept of temperature is a macroscopic concept that is established for the first time in a macroscopic object that is an aggregate of many atomic molecules. The huge number of particles is the premise of treating thermodynamics statistically and is the basis for applying statistical laws.

   Furthermore, the relative classical fluctuation of the physical quantity f, $\sqrt{\overline{(\Delta f)^2}}/f$, is proportional to $1/\sqrt{N}$ (N is the number of particles). Therefore, it decreases rapidly as the number of particles increases. If the fluctuation can be neglected in this way, it is possible to describe the system with an average value, and the actual thermodynamic quantity is a statistical average.

   On the other hand, energy, *for example*, is quantized in a bound state in quantum mechanics, and can only take a discrete value so that the energy changes discontinuously. However, when the number of particles constituting an object becomes enormous, the interval between the discrete energy values becomes extremely small and energy can be thought of as changing continuously in practice, and in this sense, can be described in classical physics. The fact that a macroscopic object is made up of an enormous number of particles is a prerequisite for the establishment of the thermodynamic concept and is the source of classical determinacy and continuity of thermodynamic quantities.

   (3) **Source of irreversibility**

   The law of increase of entropy in thermodynamics is an expression of the irreversibility. This law is interpreted as stating that the system always tends to transit from a low-probability state to a high-probability state by statistical mechanics. Microscopic reversibility holds for the dynamic process of each particle. Thus, the source of irreversibility lies not in the dynamical nature of statistical mechanics, but in its statistical nature. In other words, the irreversibility increases as the number of particles increases.

(4) **Virtually isolated system**
A macroscopic object is made up of a huge number of particles. Therefore, the proportion of particles near the surface is extremely small compared to the total number of particles of the macroscopic object. Therefore, the energy of the interaction of the macroscopic object with the surroundings is smaller than the internal energy. For this reason, it can be said that macroscopic objects are almost isolated. But here is something to be emphasized. That is, it is only possible to say that macroscopic objects are almost isolated for not too long a time. The fact that this macroscopic object is almost isolated has a very important meaning, as already mentioned. That is, we know that the majority of physically interesting quantities are additive. This situation is a consequence of the fact that individual parts of an object are macroscopic and therefore almost isolated, and the value of such a quantity for an object is equal to the sum of this quantity for an individual macroscopic part.

Considering the above, we will see how important it is to be macroscopic in thermodynamics. The concept of thermodynamics itself is established only because the object is macroscopic. The fact that

(1) the fluctuation of a certain physical quantity is small, and therefore, the state of the system can be described by the thermodynamic quantity as its average value,
(2) it changes continuously, and
(3) it is additive,

is all guaranteed by the condition that all the objects are macroscopic.

Therefore, we must remember that such a thermodynamic quantity, even if we apply it to a molecule, is not really a quantity for a single molecule, but is in some sense an average, a formal assignment of a macroscopic quantity to one. Furthermore, the most characteristic feature of thermodynamics, irreversibility (the law of increase of entropy in isolated systems), has its origin in the macroscopic nature of matter.

2. **Science describing equilibrium states, or static states.**

It is clear that thermodynamics is essentially related with an equilibrium state, if we think about the fact, as already mentioned, that temperature, *for example*, can be defined only in an equilibrium state. About entropy, equilibrium state is also assumed basically, as can be seen from the two ways of definition. If the whole system is not in equilibrium, it is divided into parts and entropy is defined for the subsystem in which equilibrium is established. Even in non-equilibrium thermodynamics (*or*, called "irreversible thermodynamics") at the present stage, local (partial) equilibrium is assumed. Therefore, systems with large deviations from their equilibrium states are not originally taken into account. For this reason, thermodynamics is effective only for systems in equilibrium, and is completely powerless for systems in non-equilibrium. Therefore, thermodynamics cannot be used for phenomena that are largely out of equilibrium (*e.g.*, chemical reactions with activation energy) or changing processes.

Thermodynamics is not a theoretical system suitable for dealing with change, as follows. As noted above, thermodynamic quantities are defined for an equilibrium state. The basic quantities of thermodynamics, such as internal energy and entropy, are state quantities. That is, it does not depend on the process of change, the path of the change, the rate of the change on the way, or anything else, so such a change in the thermodynamic state quantity depends only on the initial and the final states, and not on the process and the state on the way involved at all. This suggests that even when dealing with change, it is a completely non-causal treatment.

The thermodynamic "present" state has nothing to do with the "past" state and it does not remember any history and past of itself. As far as such a state quantity is concerned, only the states that are unrelated to each other are important, and the connection between the states and the rate of transition to another state are completely out of interest. Thus, the relationship between the two states of thermodynamics is a completely separated relation between the two states, and a causal relationship between them should not be admitted, and therefore nor the rate of change between them become a subject of discussion. Thus, the concept of time is not involved in thermodynamics. Such a situation means that the world of thermodynamics is essentially a science of equilibrium states or the static world, and thermodynamics is essentially inappropriate for dealing with changes such as chemical reaction rate or motion itself.

3. **Usefulness for necessary conditions, powerless for sufficient conditions**

Thermodynamics is useful only in the formulation of the necessary condition but is powerless in the setting of the sufficient condition. For example, let us assume the case that liquid water is formed from 1 atm of hydrogen and 1 atm of oxygen as

$$H_2(\text{gas}) + \frac{1}{2}O_2(\text{gas}) = H_2O(\text{liq}) : \Delta G^\circ_{298} = -56{,}690 \text{ cal/mol}$$

The Gibbs free energy decrease in this reaction is 56,690 cal/mol at 25 °C [7, Chap. 8]. It is generally said that the driving force for this reaction is a decrease in the Gibbs free energy. However, the following points should be noted. Decrease in Gibbs free energy means decrease in free energy when this reaction actually occurs and does not mean this reaction occurs because of the decrease in Gibbs free energy. Reactions that will cause a decrease in Gibbs free energy may or may not occur in reality.

In this case, the reaction does not occur when 1 atm of hydrogen and 1 atm of oxygen are mixed and kept at 25 °C. It responds instantly if you fly an electric spark. In general, thermodynamically impossible changes do not occur spontaneously. Also, possible changes due to thermodynamics may or may not occur spontaneously. In short, thermodynamics is not effective for setting sufficient conditions but responsible only to the necessary conditions for changes. In the above example, the decrease in Gibbs free energy is the result of this reaction and not the cause of this reaction. In this sense, the statement that the decrease in Gibbs free energy is a driving force of the reaction is misleading and should not be used.

4. **Other miscellaneous thermodynamic features**

There are some other features of thermodynamics other than those mentioned above. Since it seems unnecessary to explain in detail, we only list them as follows:

(1) The discovery of the laws is empirical, based on human failures. Therefore, it is not clear to what extent these laws can be applied when considering each stage and each process of natural hierarchical and historical development.

That is,

(2) Although thermodynamics is considered to hold quite generally, the general basis for its scope is not always clear.
(3) Thermodynamics uses tricky mathematics, which makes it difficult to find a way to formulate a problem that leads to the desired result.
(4) There is no reliable way to come up with a cycle that will give us the information we need. This makes thermodynamics even more difficult to use. These four difficulties described in this section are quite superficial, unlike those described in the previous section. Therefore, it is thought that it can be overcome if we become familiar with thermodynamics.

1.8.4 Quantum Mechanics, Classical Mechanics, and Thermodynamics—Statistical Physics

The characteristics of quantum mechanics, classical mechanics, and thermodynamics described above can be summarized as follows in relation to reaction kinetics.

First, as for quantum mechanics,

(1) It individually describes the dynamic phenomena of fine particles such that the dynamic quantity of one particle is almost the same as the Planck constant.
(2) The concept of time is used in the wave equation, which is a basic equation, and represents the change of a state with time. The wave equation is symmetric with respect to the reversal of time, so the phenomenon described is reversible.
(3) When the Hamiltonian explicitly includes time, energy is generally not conserved. Therefore, here is no stationary state. If the Hamiltonian does not include time, *that is*, if the system is isolated, the energy is conserved, and its value is definite and thus a stationary state is achieved.
(4) Quantum mechanics, which consists of descriptions using wave function and observations, is a theoretical system with a stochastic character as a whole.

(5) The uncertainty principle that reflects the particle and wave duality of matter is established, and the quantities conjugated to each other cannot have definite and constant values at the same time.
(6) In the bound state, the eigenvalues of the Hamiltonian are quantized and can only take discrete values.

Next, as for classical mechanics,

(1) It individually describes the dynamic phenomena of a particle where the dynamic quantity of one particle is much larger than the Planck's constant.
(2) The concept of time is used in the equation of motion, which is the basic equation, and represents the change of the state with time. The equation of motion is symmetric with respect to the reversal of time, so the phenomenon described is reversible.
(3) The law of conservation of energy is derived from homogeneity in time, the law of conservation of momentum is derived from homogeneity in space, and the law of conservation of angular momentum is derived from isotropy in space.
(4) Classical mechanics is a deterministic theoretical system in the sense that all the physical quantities it treats are definitive, and the state of the future system is uniquely determined when an initial value is given.
(5) All materials have the property of either particles or waves, and they do not behave together. Therefore, the uncertainty principle does not hold.
(6) All the physical quantities to be dealt with change continuously, and do not change discontinuously.

Further, as for thermodynamics,

(1) This is a theoretical system that describes the temperature-related phenomena of a system consisting of a huge number of particles from a macroscopic standpoint.
(2) The three laws of thermodynamics do not include the concept of time, and therefore cannot describe the rate of change, but can deal with the direction of change and irreversible change by using state quantities.
(3) The law of conservation of energy holds for isolated systems.
(4) Thermodynamics targets a system consisting of an enormous number of particles, and there are no particular restrictions on the size of the dynamic quantity of one particle. Therefore, the theoretical system has a built-in stochastic/probabilistic property in the dual sense of quantum statistics and classical statistics in itself.
(5) Thermodynamics seems to have a quantum mechanical character and a classical mechanical character in essence, but the physical quantity to be dealt with is semi-definite and the change will be semi-continuous, depending on the macroscopic character of the object.
(6) Thermodynamics ultimately holds for isolated equilibrium systems and is essentially static. Therefore, when dealing with change, we are dealing with two equilibrium states, and we cannot describe the intermediate states of the change. Also, the logic merely formulates a necessary condition, not a sufficient condition.

Comparing the characteristics of quantum mechanics, classical mechanics, and thermodynamics described above, we can see that they are complementary to each other. Moreover, there is a relationship quite interesting among them, as follows:

Quantum Mechanics and Classical Mechanics

Let us talk about the relationship between quantum mechanics and classical mechanics. As is clear from the comparison of the characteristics of these two, quantum mechanics holds for fine particles in which the magnitude of the dynamic quantity of one particle is almost the same as Planck's constant, whereas classical mechanics holds for a particle whose dynamic quantity is much larger than Planck's constant. Therefore, classical mechanics can be formally considered to be the limit case where Planck's constant $h \to 0$ in quantum mechanics. Bohr called such a situation the "correspondence principle" and expressed it as follows. In other words, in the limit where the action of interest is large enough to ignore individual quanta, fundamentally statistical stochastic description of quantum phenomena is a rational generalization of classical mechanical description. Alternatively, the laws of quantum mechanics must agree with the classical mechanics formula in the classical limit where the quantum number increases.

Quantum mechanics and classical mechanics not only agree in their limit ($h \to 0$ or $\Delta n \ll n(n :$ quantum number)), but also have a closer relationship, as is clear from the Ehrenfest theorem described below.

Now, let us consider the change of the mean value of the position of one particle with time. If ψ is a normalized wave function, the mean value of the x coordinate is given by

$$\langle x\rangle = \int\int_{-\infty}^{\infty}\int \psi^* x\psi \mathrm{d}x\mathrm{d}y\mathrm{d}z \equiv (\psi, x\psi) \tag{1.15}$$

When both sides are differentiated by time t and the wave equation is applied, the Hermitian property of Hamiltonian gives

$$\begin{aligned}\frac{\mathrm{d}}{\mathrm{d}t}\langle x\rangle &= \left(\frac{\partial\psi}{\partial x}, x\psi\right) + \left(\psi, x\frac{\partial\psi}{\partial t}\right) = \frac{i}{\hbar}\{(H\psi, x\psi) - (\psi, xH\psi)\}\\ &= \frac{i}{\hbar}\{(\psi, Hx\psi) - (\psi, xH\psi)\} = \left\langle\frac{i}{\hbar}(Hx - xH)\right\rangle\end{aligned} \tag{1.16}$$

For one particle, Hamiltonian is

$$H = -\frac{\hbar^2}{2m}\Delta + V(x) \tag{1.17}$$

If you put this expression in H,

$$\frac{i}{\hbar}(Hx - xH) = \frac{\hbar}{mi}\frac{\partial}{\partial x} \tag{1.18}$$

Considering

$$p_x \equiv \frac{\hbar}{i}\frac{\partial}{\partial x} \tag{1.19}$$

and you will obtain

$$m\frac{\mathrm{d}}{\mathrm{d}t}\langle x\rangle = \langle p_x\rangle \tag{1.20}$$

A similar equation is obtained for the y and z coordinates. In the case of y, p_y is used instead of p_x, and in the case of z, p_z is used. Therefore,

$$m\frac{\mathrm{d}}{\mathrm{d}t}\langle \boldsymbol{x}\rangle = \langle \boldsymbol{p}\rangle \tag{1.21}$$

Thus, the same equation holds for the mean value as in classical mechanics. Differentiating both sides of Eq. (1.20) with t again,

$$\begin{aligned} m\frac{\mathrm{d}^2}{\mathrm{d}t^2}\langle x\rangle &= \frac{\mathrm{d}}{\mathrm{d}t}\langle p_x\rangle = \frac{\hbar}{i}\frac{\mathrm{d}}{\mathrm{d}t}\left(\psi, \frac{\partial}{\partial x}\psi\right) = \frac{\hbar}{i}\left\{\left(\frac{\partial\psi}{\partial t}, \frac{\partial}{\partial x}\psi\right) + \left(\psi, \frac{\partial}{\partial t}\frac{\partial}{\partial x}\psi\right)\right\} \\ &= \left(H\psi, \frac{\partial}{\partial x}\psi\right) - \left(\psi, \frac{\partial}{\partial x}H\psi\right) = \left[\psi, \left(H\frac{\partial}{\partial x} - \frac{\partial}{\partial x}H\right)\psi\right] \end{aligned} \tag{1.22}$$

If we put Eq. (1.17) into Hamiltonian H,

$$H\frac{\partial}{\partial x} - \frac{\partial}{\partial x}H = -\frac{\partial V}{\partial x} \tag{1.23}$$

In the cases of $\langle y\rangle$ and $\langle z\rangle$, partial derivatives of y and z appear, respectively. Therefore,

$$m\frac{\mathrm{d}^2}{\mathrm{d}t^2}\langle \boldsymbol{x}\rangle = -\langle \mathrm{grad}\ V\rangle \tag{1.24}$$

is obtained. Since $-\mathrm{grad}\,V(\boldsymbol{x})$ represents the classical mechanical force acting on this particle at point $\boldsymbol{x}$, the above equation is the classical mechanics (Newton) equation of motion for the quantum mechanical mean values.

This result is called Ehrenfest theorem. It must be pointed out that this theorem holds, as is clear from the derivation method, in an isolated state, *that is*, when no force other than the equation of motion shows. The ability to derive the equation of motion of classical mechanics as an average value in quantum mechanics means that the quantum mechanical stochastic law is more fundamental than the classical mechanical deterministic law, and the classical mechanical deterministic law approximates the quantum mechanical stochastic law. In other words, motions with quantum mechanical fluctuations should originally be described in quantum mechanics. If motions were described in classical mechanics, it is equivalent to that they were described in terms of averages, ignoring the fluctuations. Such an approximation by the average value is a good approximation when the Planck constant can be regarded as zero and therefore the fluctuation is small.

Let us consider the relationship between quantum or classical mechanics and thermodynamics. In short, this relationship is that between an individual description and a collective description. Both are related to each other by Gibbs distribution as described below.

Quantum Mechanics and Thermodynamics

We consider the relationship between quantum mechanics and thermodynamics. For this purpose, we will consider a macroscopic object with a huge number of fine particles, the motion of which is described by quantum mechanics. Our aim is to find the probability w_n that such a system of macroscopic objects will be found in a particular quantum state of energy E_n [8, Chap. 3].

Let us now separate the object of interest from the isolated system and consider this isolated system to be composed of both parts of the object of interest and the remaining part. Let us call the former the "object" and the latter the "environment" for the object. Let E, $d\Gamma$ and E', $d\Gamma'$ be the energies and the number of quantum states of the object and the environment, respectively. $E^{(0)}$ is a predetermined energy value of the isolated system. Since the sum of the energies of the object and the environment, $E + E'$, must be equal to this value $E^{(0)}$, the energy distribution of this isolated system (called the micro-canonical ensemble) is given by

$$\mathrm{d}w = \text{const} \cdot \delta\left(E + E' - E^{(0)}\right)\mathrm{d}\Gamma\mathrm{d}\Gamma' \tag{1.25}$$

We are interested in the probability that the energy of the object is E_n and not interested in the energy of an environment. So, let us assume that the environment is in a certain macroscopically described state. The probability w_n that the energy of the object is E_n is given by Eq. (1.26), which is obtained by integrating Eq. (1.25) with respect to $\mathrm{d}\Gamma'$, while $\mathrm{d}\Gamma$ is set to 1 and $E = E_n$.

$$w_n = \text{const} \cdot \int \delta\left(E_n + E' - E^{(0)}\right)\mathrm{d}\Gamma' \tag{1.26}$$

Suppose that the total number of quantum states of the environment whose energy is smaller than or equal to E' is $\Gamma'\left(E'\right)$. Because the integrand in the above equation depends only on E', $\mathrm{d}\Gamma'$ can be written as

$$\mathrm{d}\Gamma' = \frac{\mathrm{d}\Gamma'\left(E'\right)}{\mathrm{d}E'}\mathrm{d}E' \tag{1.27}$$

and this integration can be converted into the integration of dE'.

Let us replace derivative $d\Gamma'/dE'$ with a ratio like

$$\frac{d\Gamma'}{dE'} = \frac{\exp[S'(E')/k]}{\Delta E'} \tag{1.28}$$

where $\Delta E'$ is the interval between the energy values of the environment corresponding to the quantum state interval $\Delta\Gamma'$, and $S'(E')$ is the entropy of the environment as a function of energy.

Thus,

$$w_n = \text{const} \cdot \int \frac{\exp[S'/k]}{\Delta E'} \delta(E_n + E' - E^{(0)}) dE' \tag{1.29}$$

Due to the existence of the δ-function, this integral simply replaces E' with $E^{(0)} - E_n$, we obtain

$$w_n = \text{const} \cdot \left(\frac{\exp[S'/k]}{\Delta E'} \right)_{E'=E^{(0)}-E_n} \tag{1.30}$$

Now, let us assume that the energy E_n is smaller than $E^{(0)}$ because the object is smaller than the environment. For small changes in E', $\Delta E'$ changes relatively little. Therefore, we can set $E' = E^{(0)}$, and if we replace it in this way, ΔE becomes a constant irrelevant to E_n. For $\exp[S'/k]$, by expanding $S'(E^{(0)} - E_n)$ by the width of E_n and taking up the first-order term, we obtain

$$S'(E^{(0)} - E_n) = S'(E^{(0)}) - E_n \frac{dS'(E^{(0)})}{dE^{(0)}} \tag{1.31}$$

By the way, the derivative of the entropy S' by energy is nothing more than $1/T$, where T can be considered as the temperature of the system, because the system is now in equilibrium, the temperatures of the object and the environment are equal. Thus, for all w_n we obtain

$$w_n = A\exp\left(-\frac{E_n}{kT}\right) \tag{1.32}$$

where A is a normalization constant irrelevant to E_n.

This is one of the most important formulas in statistical physics, which determines the statistical distribution of any macroscopic object that forms a relatively small part of a large isolated system. The distribution in Eq. (1.32) is called the Gibbs distribution or the canonical distribution. The normalization constant, A, is determined by the condition that $\sum_n w_n = 1$ and

$$\frac{1}{A} = \sum_n \exp\left(-\frac{E_n}{kT}\right) \tag{1.33}$$

The mean value of any physical quantity f which characterizes the given object can be calculated with the help of Gibbs distribution:

$$\overline{f} = \sum_n w_n f_{nn} = \frac{\sum_n f_{nn} \exp(-\frac{E_n}{kT})}{\sum_n \exp(-\frac{E_n}{kT})} \tag{1.34}$$

where $f_{nm} = \int \psi_n^* f \psi_m \mathrm{d}q$.

At this point, we can discuss the relationship between quantum mechanics and thermodynamics. For example, the thermodynamic energy E is the mean value of the quantum mechanical energy E_n Therefore, according to Eq. (1.34)

$$E \equiv \overline{E}_n = \frac{\sum_n E_n \exp(-\frac{E_n}{kT})}{\sum_n \exp(-\frac{E_n}{kT})} \tag{1.35}$$

Entropy S is given as

$$S = -k\overline{\ln w_n} = -k\ln A + \frac{\overline{E}_n}{T} = k\ln A + \frac{E}{T} \tag{1.36}$$

The Helmholtz free energy F is $F = E - TS$, so $F = kT\ln A$. Using this, the Gibbs distribution becomes

$$w_n = \exp\left(\frac{F - E_n}{kT}\right) \tag{1.37}$$

Therefore,

$$\sum_n w_n = \exp\left(\frac{F}{kT}\right) \sum_n \exp\left(-\frac{E_n}{kT}\right) = 1 \tag{1.38}$$

Using this equation,

$$F = -kT\ln \sum_n \exp\left(-\frac{E_n}{kT}\right) \tag{1.39}$$

These formulas are the basis for applying the Gibbs distribution to thermodynamics. These clearly show the relationship between quantum mechanical quantities and thermodynamic quantities. In other words, the thermodynamic quantity is a quantum statistical average weighted by Gibbs distribution. The fact that thermodynamic quantities can be derived from quantum mechanical quantities as their mean values in this way means that quantum mechanical quantities and their theoretical systems are more fundamental than thermodynamic quantities and their theoretical systems. Thermodynamics is a macroscopic approximation of quantum mechanics.

In other words, if quantum phenomena, which should be described in quantum mechanics, are described thermodynamically, it is equivalent to an approximate description with a mean value by Gibbs distribution. A description with such an average is a good approximation when the fluctuation is so small that a description ignoring the fluctuation is possible. The relation between quantum mechanics and thermodynamics is a remarkable contrast to the relation between quantum mechanics and the classical mechanics. The average in the latter is performed purely by quantum mechanics, while the average in the former is performed with the help of not purely the quantum mechanics but that of statistics (Gibbs distribution).

Classical Mechanics and Thermodynamics

The relationship between classical mechanics and thermodynamics will be discussed. This can be attained based on a slight modification of the Gibbs distribution formula which links quantum mechanics and thermodynamics. In the Gibbs distribution formula (1.32), the energy E_n was quantized and changed discretely. In classical mechanics, Planck's constant h is considered to be zero, and the gap between discrete energies is regarded as zero. *That is*, the energy is considered to change continuously. Thus, the Gibbs distribution in classical statistics is

$$\rho(p,q) = A\exp\left[-\frac{E(p,q)}{kT}\right] \tag{1.40}$$

Here, p and q represent the momentum and coordinates. A is determined by the following equation for normalization:

$$\int \rho \mathrm{d}p\mathrm{d}q = A\int \exp\left[-\frac{E(p,q)}{kT}\right]\mathrm{d}p\mathrm{d}q = 1 \tag{1.41}$$

In this case, the mean value of physical quantity, f, is given by

$$\overline{f} = \int f\exp\left[-\frac{E(p,q)}{kT}\right]\mathrm{d}p\mathrm{d}q \Big/ \int \exp\left[-\frac{E(p,q)}{kT}\right]\mathrm{d}p\mathrm{d}q \tag{1.42}$$

Therefore, thermodynamic energy, entropy, and Helmholtz's free energy can be obtained as well as before:

$$E = \int E(p,q)\exp\left[-\frac{E(p,q)}{kT}\right]\mathrm{d}p\mathrm{d}q \Big/ \int \exp\left[-\frac{E(p,q)}{kT}\right]\mathrm{d}p\mathrm{d}q \tag{1.43}$$

$$S = -k\overline{\ln[(2\pi\hbar)^s\rho]} = -k\int \rho\ln[(2\pi\hbar)^s\rho]\mathrm{d}p\mathrm{d}q \tag{1.44}$$

$$F = -kT\ln\int{}' \exp\left[-\frac{E(p,q)}{kT}\right]\mathrm{d}\Gamma = -kT\ln\int{}' \left[\exp\left[-\frac{E(p,q)}{kT}\right]/(2\pi\hbar)^s\right]\mathrm{d}p\mathrm{d}q \tag{1.45}$$

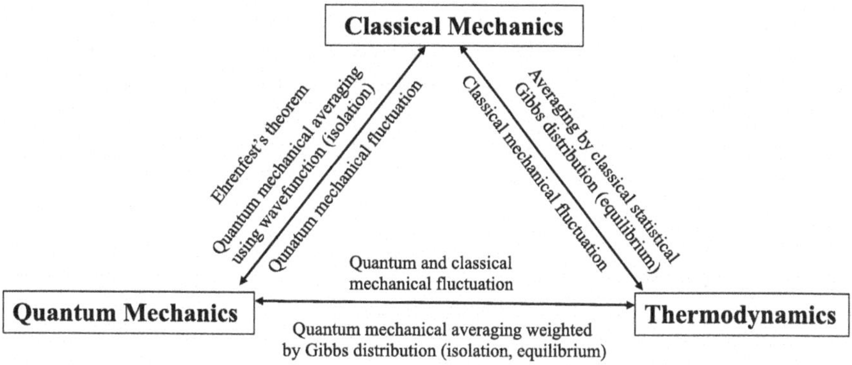

Fig. 1.3 Relationship between quantum mechanics, classical mechanics, and thermodynamics

Here, the dash of integration means that integration will be conducted only in the phase space corresponding to a physically different state, and s is the degree of freedom.

Those formulae are the basis for applying the Gibbs distribution in classical statistics to thermodynamics. This Gibbs distribution links classical mechanics with thermodynamics, and the thermodynamic quantity is the average value of the classical mechanical quantity by the classical statistical Gibbs distribution. In case what is supposed to be described in classical mechanics is described in thermodynamics, it is an approximate description based on the average values. Note that the description with the average is a good approximation when the fluctuation is small and can be ignored.

Thus, the relationships have been described above between quantum mechanics and classical (Newtonian) mechanics, between quantum mechanics and thermodynamics, and between classical mechanics and thermodynamics. They are shown in Fig. 1.3.

1.8.5 Quantum and Classical Phenomena—Criteria of Distinction

In the previous section, we considered the logic of quantum mechanics and classical mechanics and their characteristics and relationships. Two theoretical systems are separately used, depending on quantum and classical phenomena, respectively. Thus, the question comes arise to describe a concrete phenomenon, whether we should choose quantum mechanical description or classical mechanical description. In order to find out the solution to this question, we must judge whether the phenomenon is fundamentally derived from quantum mechanical or classical basis. In this section, we will take a closer look at the criterion of the logic basis.

Terminology of "classical" is used in a variety of different ways. This means that the word of "classical" is not really well-defined. *For example*, the meaning of the word "classical" for the classical music or the classical art has a nuance that it is still valid in some fields but what we do not apply today. In this sense, classical physics may be called "classical" and the content and structure of classical physics does not mean out of date. The expression of "classical mechanics" is used with two meanings. This terminology is used to express the non-relativistic physics (or it does not consider relativistic effects), as opposed to the relativistic physics that deals with the phenomenon when the velocity of motion of the particle approaches the speed of light. On the other hand, the terminology is also used that the physical object can be described, ignoring the quantum effect. After all, classical mechanics can be said to be an extremum that holds for the limit of the world where the velocity of light is regarded as $c \to \infty$ and Planck's constant $h \to 0$, that is, non-relativistic non-quantum mechanics. In the following sections, this terminology will be used to express the counterpart of quantum mechanics.

The first lemma: a particle is classical if its magnitude of dynamical quantities (mass, momentum, energy, angular momentum, etc.) is sufficiently large compared to Planck's constant h. In case they are about the same in magnitude, we should deal the particle with quantum mechanics. For instance, let us consider the case where a particle with a large mass m moves. In this case, the momentum of the particle p is expressed as mv, and the uncertainty relation $\Delta x \cdot \Delta p = m(\Delta x \cdot \Delta v) \cong \hbar$ becomes

$$\Delta x \cdot \Delta v = \frac{\hbar}{m} \tag{1.46}$$

Accordingly, the quantum mechanical uncertainty can be ignored if m is large. When the mass m of the traveling particle is larger than h, the gap between discrete eigenvalues becomes so small that it can be ignored. As far as the motion of each particle is concerned, the criterion of the behavior is thus subject to the condition whether its dynamical quantity is sufficiently large compared to h or not.

The second lemma: If the phenomena are determined by the exchange of several quanta in energy, the phenomenon should refer to a quantum mechanical principle. On the contrary, if the exchange of several quanta gives no influence to the phenomenon, we describe the phenomenon as a classical event. If a microscopic object receiving a significant influence from the exchange of several quanta in the system, the phenomenon is quantum mechanical. On the other hand, if it is a large object and the exchange of several quanta does not matter, the phenomenon can be said to be classic. However, we should note in case amplification by chain reaction of critical quantum processes raises the phenomenon in a classically observable stage that the event is essentially quantum mechanical. The photoelectric and ionization effects by light, the Compton, Zeeman effects, etc. are typical examples of such quantum phenomena. The energy spectrum lines are also purely quantum mechanical data. Thermodynamics deals with a system consisting of a huge number of particles. However, it is essentially important to distinguish whether the concerned phenomenon is based on classical or quantum effect.

The third lemma: The phenomenon is quantum mechanical, if the potential appreciably changes in a range narrower than wavelength λ, otherwise it is classical. This lemma may be easily accepted, considering the de Broglie relation stating that any particle is accompanied by a wave property with a wavelength λ given by

$$\lambda = \frac{h}{p} \tag{1.47}$$

where p is the momentum of the particle. When the wavelength λ is calculated using this formula, for an object with the mass of 1 g traveling with a velocity of 1 cm/s, $\lambda = h/p = 6.63 \times 10^{-19}$ Å, the quantum effect does not explicitly appear because the wavelength is too short. However, supposing an example that the particles moving under the Brownian effect, small particles having a diameter of about 1 μm and a mass $m \cong 10^{-12}$ g move under the thermodynamic equilibrium state at the normal temperature, we easily estimate its kinetic energy $3kT/2$ as $\sim 0.4 \times 10^{-13}$ erg, and the average wavelength as

$$\lambda = \frac{h}{p} = \frac{h}{\sqrt{3mkT}} \cong 5 \times 10^{-6}\ \text{Å} \tag{1.48}$$

For the same energy, a helium atom has a wavelength of $\lambda \cong 0.9$ Å, a neutron has $\lambda \cong 1.8$ Å, and an electron has $\lambda \cong 0.9$ Å. Thus, we can expect that the quantum effect appears when these particles move in the fluctuating field where the potential energy changes with distance in the order of Angstrom. Eventually, diffraction phenomena based on the wave property of the electrons, the neutrons, the helium atoms, and the hydrogen molecules with solid crystals were experimentally confirmed. As clear from the above speculation, whether a particle is a quantum or a classical one is not absolute, but rather relative; it behaves either in a quantum or in a classical manner, depending on the field of particle's location and the way of interaction.

1.8.6 Classical and Quantum Fluctuations—Criteria of Distinction

In the previous section, we considered quantum mechanics, classical mechanics, and thermodynamics, and further explored the interrelationship of these three. We clarified that the node between quantum mechanics and classical mechanics is set by quantum mechanical averaging using wave functions, and that the linkage between quantum mechanics and thermodynamics, and between classical mechanics and thermodynamics is accomplished in terms of the Gibbs statistical distribution. Thus, in short, these three scientific systems are connected by the average operation. We should pay a close attention to such averaging operations because the physical quantity being considered does not take a definite value but involves fluctuations. In fact, the terminology of "average" and "fluctuation" includes two types of concepts: the average accompanying classical fluctuation and the average accompanying quantum mechanical fluctuation. Both characteristics are quite different with each other. The detail is now described below.

Classical Fluctuation and Averaging

We consider the characteristics of classical fluctuation and the average. Now suppose that a macroscopic object consists of a huge number of particles. The behavior of each particle is described by classical mechanics or quantum mechanics if it comes from the classical origin or quantum origin, respectively. Therefore, classical equations of motion or wave equations must be established for each particle and solved, to discuss the state of a system consisting of an enormous number of particles. Generally speaking, the equation including a huge number of parameters is hard to be solved in practice.

For example, it is difficult to integrate the equation of motion of classical mechanics in a general form. Even if it could be integrated in a general form, it is generally impossible to put the initial conditions for the velocity and coordinates of the particle in the solution because of the time and the amount of paper required for it. On the other hand, the integral of an enormous number of wave equations is a much less probable problem than the integral of the equation of motion in classical mechanics. Even if a general solution of the wave equation could be found by chance, it would be implausible to assign and write it down a special solution that satisfies the given concrete conditions for a problem, characterized by a certain value for each one of various and enormous quantum numbers.

When dealing with a system consisting of an enormous number of particles, we thus usually rely on a statistical approximation method without having accurate knowledge about the dynamic state of the system. As a result, the description of the system becomes stochastic. It should be noted that the source of stochasticity is not in the properties of the system itself, but in the description method we have adopted, or in the description method of statistical approximation. It is a stochasticity and fluctuation that would not have been brought about when the general solution of the equation could be decided without the saving of time and paper. In short, the classical fluctuations are generated by the adoption of approximate methods in statics, instead of rigorous calculations due to difficulty to treat a huge number of coordinates of particles, despite that the complete information about the system is in principle possible for us to obtain. Thus, the classical fluctuation reflects the imperfection of knowledge and information. It is only an apparent fluctuation or apparent stochasticity, caused by our ignorance of nature of an object that is in fact definite.

We next show that the relative fluctuation of physical quantity with lack of information rapidly decreases as the size of the object (number of particles) increases. Now let f be an additive quantity. As already pointed out, macroscopic objects can be regarded as almost isolated systems for not long periods of time, so the physical quantities are additive. Now, we consider the object consisting of a large number (N) of components with nearly similar sizes. Then,

$$f = \sum_{i=1}^{N} f_i \tag{1.49}$$

where f_i is the quantity of i-th component. The equation also holds for the mean values as

$$\overline{f} = \sum_{i=1}^{N} \overline{f}_i \tag{1.50}$$

Thus, it is clear that $\overline{f}$ increases approximately in proportion to N as the number of components increases. The mean square of the fluctuation of the quantity f is

$$\overline{(\Delta f)^2} = \overline{\left(\sum_i \Delta f_i\right)^2} \tag{1.51}$$

Since the components of the object are statistically independent for each other, the average of the products is

$$\overline{\Delta f_i \cdot \Delta f_k} = \overline{\Delta f_i} \cdot \overline{\Delta f_k} = 0 \tag{1.52}$$

because each $\overline{\Delta f_i} = 0$. Therefore,

$$\overline{(\Delta f)^2} = \sum_{i=1}^{N} \overline{(\Delta f_i)^2} \tag{1.53}$$

This shows that the mean square $\overline{(\Delta f)^2}$ becomes large in proportion to N as N increases. Therefore, the relative fluctuation is inversely proportional to $\sqrt{N}$, then

$$\frac{\sqrt{\overline{(\Delta f)^2}}}{\overline{f}} \propto \frac{1}{\sqrt{N}} \tag{1.54}$$

If we divide a homogeneous object into smaller parts of a fixed size, the number of such parts is clearly proportional to the total number of particles (molecules). Thus, the result obtained above can be formulated as follows. The relative fluctuation of any additive quantity f decreases in inverse proportion to the square root of the number of particles contained in the object. Therefore, if the number of particles is large enough, the quantity f itself can refer to be constant over time and equal to its average value.

The classical fluctuation arises from the fact that the system consists of an enormous number of particles, and we had virtually no knowledge of all of them. Therefore, if the number of classical particles is only one and the equation of motion is solved and then the knowledge is perfect, the classical fluctuation does not arise as described above, and therefore does not matter. Although the fluctuation also appears in quantum mechanics, the "quantum mechanical fluctuation" possesses completely different characteristics.

Quantum Mechanical Fluctuation

The quantum mechanical fluctuation appears for one particle in case that the classical fluctuation is literally unexpected. This situation comes from the fact that the particle is subject to a quantum mechanical rule, that is, the magnitude in dynamical quantity of one particle is almost the same size as the Planck constant, and the particle–wave duality is remarkable in this scale. The fact that a state of which dynamical quantity cannot have definite values can be realized is one of the most characteristic features of quantum mechanics, which comes from the principle of superposition of wave functions. It should be noted here that the source of quantum fluctuations and stochasticity is different from that of classical fluctuations. The quantum fluctuation is derived from the nature of the object itself and the particle–wave duality of matter. As was already mentioned, the quantum fluctuation satisfies the uncertainty relation.

So far, we have considered the nature of two types of fluctuations, classical and quantum mechanical fluctuations. Classical statistics covers a huge number of classical particles. Only the classical fluctuations are assumed in this case. On the other hand, quantum statistics targets a large number of quantum mechanical particles. Thus, both classical and quantum mechanical fluctuations may appear simultaneously.

Therefore, the averaging operation in quantum statistics involves the dual character of the statistical mean (mean for classical fluctuations) and the quantum mechanical mean (mean for quantum mechanical fluctuations). This situation is explicit for us if we consider the relationships between quantum mechanics and classical mechanics, between classical mechanics and thermodynamics, and between quantum mechanics and thermodynamics, as shown in Fig. 1.4, and if we remember that these relationships are connected with each other through the averaging operations. These relationships may be formally represented as shown in the following figure.

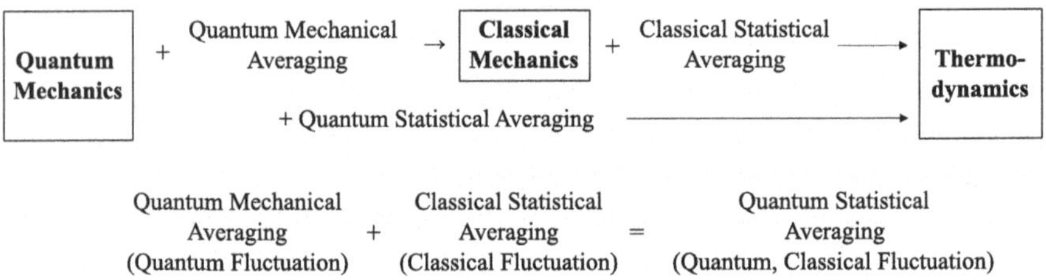

Fig. 1.4 Relationship between quantum mechanics, classical mechanics, and thermodynamics in terms of averaging operation and fluctuation

Classical and Quantum Mechanical Fluctuations

Next, let us consider the procedure to verify classical fluctuations and quantum mechanical fluctuations. Before tackling this, we start to derive the basic equations for distinguish of these two kinds of fluctuations. We need bear the fact into mind that the concept of time-derivative of physical quantities cannot be defined in quantum mechanics in the same sense as in the classical mechanics. This is due to the situation that the definition of differentiation in classical mechanics is related to the identification of the value of a physical quantity at two different but close times and in quantum mechanics, on the other hand, the quantity exhibiting a definite value at a certain moment cannot have any definite value after a period of time.

For example, in classical mechanics, a particle has definite values of coordinate and velocity at each time, but in quantum mechanics, the uncertainty principle $\Delta x \cdot \Delta p \cong \hbar$ causes the following situation. If an electron is found to have a definite coordinate as a result of measurement, then the electron does not generally have any definite velocity. Conversely, an electron having a definite velocity cannot have any definite position in space. In fact, the simultaneous determination of coordinates and velocity at any given time indicates that there is an explicitly defined trajectory. However, in the case of electrons, this presumption cannot be applied any more.

In quantum mechanics, the coordinates and velocity of an electron are conjugate quantities that cannot be precisely measured simultaneously. That is, two quantities do not have definite values at the same time. Thus, the coordinates and velocity of an electron cannot be determined at the same time. This also suggests that we cannot define the velocity as a time-derivative of the coordinates by knowing the definite values of the coordinates at two adjacent times.

In quantum mechanics, therefore, the concept of time-derivative needs to be defined differently [9, Chap. 2]. It may be natural and acceptable for us to define the time-derivative of the quantity, $\dot{f}$, as such quantity that the mean value of $\dot{f}$, $\overline{\dot{f}}$, is equal to the time derivative of the mean $\overline{f}$, $\dot{\overline{f}}$. Hence, according to this definition

$$\overline{\dot{f}} = \dot{\overline{f}} \tag{1.55}$$

Using this definition, we easily get the expression of the quantum mechanical operator corresponding to the quantity $\dot{f}$. Since

$$\overline{f} = \int \psi^* \hat{f} \psi \mathrm{d}q \tag{1.56}$$

then we have

$$\overline{\dot{f}} = \dot{\overline{f}} = \frac{\mathrm{d}}{\mathrm{d}t} \int \psi^* \hat{f} \psi \mathrm{d}q = \int \psi^* \frac{\partial \hat{f}}{\partial t} \psi \mathrm{d}q + \int \frac{\partial \psi^*}{\partial t} \hat{f} \psi \mathrm{d}q + \int \psi^* \hat{f} \frac{\partial \psi}{\partial t} \mathrm{d}q \tag{1.57}$$

The operator $\hat{f}$ depends on time like as a parameter in some cases and $\partial \hat{f}/\partial t$ is an operator obtained by differentiating $\hat{f}$ by time. Substituting Eq. (1.1) in the derivative of $\partial \psi/\partial t$ or $\partial \psi^*/\partial t$, we can obtain

$$\overline{\dot{f}} = \int \psi^* \frac{\partial \hat{f}}{\partial t} \psi \mathrm{d}q + \frac{i}{\hbar} \int \left(\hat{H}^* \psi^*\right) \hat{f} \psi \mathrm{d}q - \frac{i}{\hbar} \int \psi^* \hat{f} \left(\hat{H} \psi\right) \mathrm{d}q \tag{1.58}$$

Since the operator $\hat{H}$ is Hermitian,

$$\int \left(\hat{H}^* \psi^*\right)\left(\hat{f} \psi\right) \mathrm{d}q = \int \psi^* \hat{H} \hat{f} \psi \mathrm{d}q \tag{1.59}$$

Therefore,

$$\overline{\dot{f}} = \int \psi^* \left(\frac{\partial \hat{f}}{\partial t} + \frac{i}{\hbar} \hat{H} \hat{f} - \frac{i}{\hbar} \hat{f} \hat{H} \right) \psi \mathrm{d}q \tag{1.60}$$

On the other hand, from the definition of mean value

$$\overline{\dot{f}} = \int \psi^* \dot{f} \psi \mathrm{d}q$$

then the expression in parentheses of the integrand is clearly the operator $\hat{\dot{f}}$ to be obtained as

$$\hat{\dot{f}} = \frac{\partial \hat{f}}{\partial t} + \frac{i}{\hbar} \left(\hat{H} \hat{f} - \hat{f} \hat{H} \right) \tag{1.61}$$

However, we should note here the following. If the operator $\hat{f}$ does not depend explicitly on time, $\hat{\dot{f}}$ will be reduced to a commutator of $\hat{f}$ and $\hat{H}$, $\left(\hat{H}\hat{f} - \hat{f}\hat{H}\right)$, except for a multiplier. We shall consider another important category of physical quantities, i.e., "conserved quantities". We know that some operators do not depend on time, and they are commutative with Hamiltonians, where $\hat{\dot{f}} = 0$. Such quantities are subject to the "conserved quantities" in quantum mechanics. We have already mentioned that the time-independent Hamiltonian $\hat{H}$ is conserved and is called "energy".

If the operator $\hat{\dot{f}}$ becomes identically zero in any situations, then $\overline{\dot{f}} = \dot{\overline{f}} = 0$, that is, $\overline{f} = \text{const}$. In other words, the average value of this quantity is unchanged with time, or if the quantity f takes a definite value in a given state (i.e., if the wave function is an eigenfunction of the operator $\hat{f}$), it may be plausible to say that the same definite value is obtained at any times thereafter.

The above consideration leads us to attain another useful relation. Suppose f and g are two physical quantities, and their operators obey the commutative law like

$$\hat{f}\hat{g} - \hat{g}\hat{f} = -i\hbar\hat{c} \tag{1.62}$$

where $\hat{c}$ is an operator of a certain physical quantity c. The introduction of the Planck constant h on the right-hand side of this equation corresponds to the fact that all operators of physical quantities accordingly result in multiplication of their values, and they are commutative with each other in the classical limit (i.e., $h \to 0$). Therefore, in the quasi-classical case, the right side of Eq. (1.62) can be referred to zero to the first approximation. Thus, the operator $\hat{c}$ can be regarded as a simple product operator of the quantity c to the second approximation. Consequently we obtain

$$\hat{f}\hat{g} - \hat{g}\hat{f} = -i\hbar c \tag{1.63}$$

This equation is exactly the same as the relation between coordinates and momentum,

$$\widehat{p_x}x - x\widehat{p_x} = -i\hbar,$$

except that the quantity $\hbar c$ is included instead of the constant $\hbar$. Along the similarity with the relational expression $\Delta x \cdot \Delta p_x \cong \hbar$, it is concluded in the quasi-classical case that the uncertainty relation between quantities f and g holds like

$$\Delta f \cdot \Delta g \cong \hbar c \tag{1.64}$$

For instance, in the case that one of quantities is assigned as energy ($f \equiv H$) and the other as an operator ($\hat{g}$) which is explicitly independent of time, then we conclude that $c = \dot{g}$, according to Eq. (1.61). The uncertainty relation in the quasi-classical case is

$$\Delta E \cdot \Delta g \cong \hbar \dot{g} \tag{1.65}$$

Criteria of Distinction

Using those preparatory consideration, we shall discuss the procedure to judge which is classical fluctuation or quantum mechanical fluctuation. We hereafter assume a certain quantity x instead of the operator $\hat{g}$ in Eq. (1.63). In the quasi-classical case, the relationship between the quantum mechanical uncertainty of energy and the quantum mechanical uncertainty of a certain quantity x holds

$$\Delta E \cdot \Delta x \cong \hbar \dot{x}, \tag{1.66}$$

where $\dot{x}$ is the classical rate of change of the quantity x. Let τ be the time characterizing the rate of change of the quantity x in a non-equilibrium state. Then, $\dot{x} \sim x/\tau$, resulting in

$$\Delta E \cdot \Delta x \cong \hbar \frac{x}{\tau}. \tag{1.67}$$

Obviously, what we are saying about a definite value of quantity x is only true if such condition, $\Delta x \ll x$, is satisfied that its quantum mechanical uncertainty is small. From this,

$$\Delta E \gg \frac{\hbar}{\tau} \tag{1.68}$$

Thus, the quantum mechanical uncertainty of energy must be large compared to $\hbar/\tau$. The uncertainty of entropy in the present system is described as

$$\Delta S \gg \frac{\hbar}{\tau T} \tag{1.69}$$

If the entropy has a meaning in the classical sense, it is substantially necessary that the quantum mechanical uncertainty of entropy is smaller than one like

$$kT \gg \frac{\hbar}{\tau} \quad \text{or} \quad \tau \gg \frac{\hbar}{kT} \tag{1.70}$$

This is the condition referred to a classical fluctuation. If the temperature is too low or if the rate of change of the quantity x is too fast (τ is too small), the fluctuation cannot be regarded as a classical fluctuation, but the pure quantum mechanical feature becomes outstanding. As already mentioned, the classical description is an extremum case of the quantum description. That is, the classical limit of the quantum description ($\hbar \to 0$) matches the classical description. This is also true for fluctuations. In other words, classical fluctuations should match those in the classical limit of quantum fluctuations. Thus, the pure quantum mechanical fluctuation described above possesses a distinctive nature that cannot be replaced nor expressed by classical fluctuation.

1.8.7 Certainty and Uncertainty—Distinction and Combination

In the previous sections, we showed that there are classical and quantum fluctuations, which are independent concepts to each other. Hereafter we need to take dual meaning of classical and quantum aspects in fluctuations into account to discuss in further detail when certainties and uncertainties are considered. First, the uncertainty due to classical fluctuations stems from the fact that the description is statistically approximated when a system is composed of many particles. The fluctuation, as mentioned in the previous section, rapidly decreases as the number of particles in the system increases. In other words, we can say that the statistical description of a

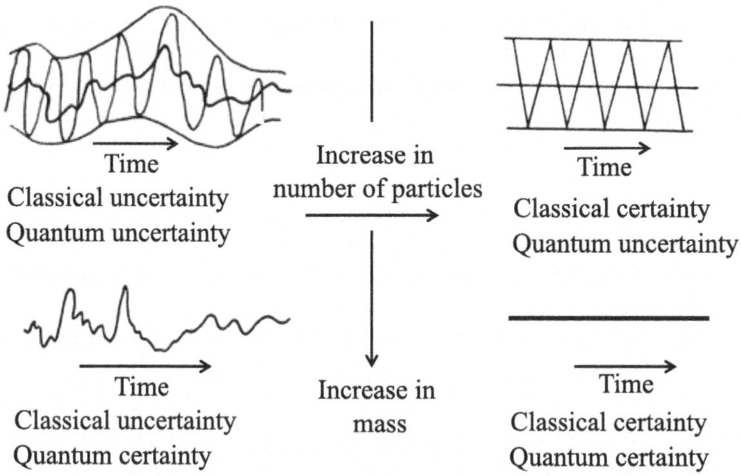

Fig. 1.5 Combinations of certainty and uncertainty in physical phenomena

system consisting of a huge number of particles is almost definite. On the other hand, quantum fluctuations are based on the fact that the particle–wave duality of matter becomes remarkable when the dynamical quantity of one particle is about the same order in magnitude as the Planck constant and satisfies the uncertainty principle. Therefore, this quantum fluctuation becomes negligibly small when the magnitude of the dynamical quantity of one particle is much larger than the Planck constant.

Considering such circumstances, the following four combinations can be considered for the combinations of certainty and uncertainty. The combination of classical certainty and quantum certainty is equivalent to describing an enormous number of classical particles by classical statistics or describing one classical particle by classical mechanics. The combination of classical uncertainty and quantum certainty is equivalent to describing a system consisting of a relatively small number of classical particles by classical statistics. Moreover, the combination of classical uncertainty and quantum certainty is equivalent to describing a system consisting of an enormous number of quantum particles by quantum statistics. Finally, the combination of classical uncertainty and quantum uncertainty corresponds to situations in which a system consisting of a relatively small number of quantum particles is described by quantum statistics. Collectively, these four kinds of combinations can be represented as shown in Fig. 1.5.

1.8.8 "Micro" and "Macro"—Connecting Factors

As already pointed out, in chemical reaction kinetics, the hierarchy involved in the cognition is the micro-hierarchy of atoms and electrons. On the other hand, the subject of cognition is human being. We humans cognize the issues occurring at the macroscopic stage in hierarchy with a spatial and temporal extent that can be identified with our sense. The relationship existing between the micro- and macro-phenomena, and key concept connecting those two phenomena emerges as an issue to be considered.

We **first** raise the Avogadro's number as a factor that connects micro- and macro-phenomena. The Avogadro number is one of the most essential concepts of moles in chemistry. The direct intuition of human being cannot reach to each atom or electron consisting of substances, and thus the human being accordingly deals with phenomena comprised a huge collection of those particles. It is through Avogadro's number, the basis of the concept of mole, that the humans can talk about the properties of individual atoms and electrons while dealing with ensemble of atoms, molecules, and ions.

The **second** essential factor is mathematical procedure of averaging for connecting micro- and macro-phenomena. If the property of object is fluctuating with time, exhibits inhomogeneity, or changes incidentally, the physical quantity under consideration does not have the same value for all particles. Thus, we describe the property of the system with an average value. As an example of such an averaging operation, we already know both the quantum mechanical average and the average by Gibbs distribution. Not only quantum mechanical averages and Gibbs distribution averages, but also statistical averages, every averaging is performed over a spatial and temporal spread to the extent as enough to allow us to identify fluctuations that we cannot directly identify with our senses. Although theoretical treatment of the macroscopic system is always based on the description of the averaged values, it should be noted that the word of "average" is sometimes inexplicitly used in such cases. In many cases, the same as the average is expressed by per unit time, per unit area, per unit volume, and so on.

In any case, the averaging exhibits the essence of statistics, playing the role of linking microscopic and macroscopic viewpoints. As already mentioned in the previous section, it is evident that thermodynamic quantities (energy, entropy, etc.) are average values when the quantities are considered in relation to quantum mechanics or classical mechanics. However, it may not be easy to notice that thermodynamic quantities themselves are average values when phenomena are considered separately from quantum mechanics or classical mechanics. Taking the fact into account that the nature has hierarchical and historical properties, and the subject of cognition is human beings, we may accept that we will inevitably cognize them as averages over a spatial and temporal extent that can be discerned by our senses when we cognize the movement of nature that belongs to a deeper level in hierarchy (space) and history (time) than the one to which we human beings belong.

When describing a system with an average value, we should remember that a system can be characterized by an average value only when the fluctuation is small. When the fluctuation is large, not only the average value but also the fluctuation itself must be considered. When the number of particles constituting the system is enormous, the fluctuation is small when the system is in equilibrium, and therefore, the characteristics of the phenomenon can be described by average values. However, when the number of particles in the system is small or the system is not in equilibrium, the fluctuation becomes large, and the approach to define the system only by the average value is not valid anymore.

1.8.9 Stationarity and Equilibrium

As for the stationary state, we have previously stated that if the Hamiltonian does not explicitly include time in the wave equation of quantum mechanics, all times are equivalent for the physical system, and the state remains unchanged over time. Here, let us consider this stationary/non-stationary state a little more. In the study of dynamical phenomena, non-stationary transport of state quantities is sometimes a problem. In a non-stationary system, the value of the state quantity at each point changes with time and causes the potential changes at the same time.

The measure of stationarity for non-stationary systems is defined by

$$K_W = \frac{W}{W + \Delta W} \tag{1.71}$$

where W is the flow of the state quantity that passes through the system, ΔW is the consumed (*or,* stored or distributed in the system) flow during the transport. If $K_W = 1(W \neq 0$ and $\Delta W = 0)$, then the system is stationary, while if $K_W < 1$ $(\Delta W \neq 0)$, then the system is non-stationary. When the system is in a stationary state, the followings can be said.

(1) At each moment, the state quantity flowing in the system is equal to the state quantity flowing out of the system.
(2) The distribution of state quantities (values of state quantities at each point) does not depend on time.
(3) The potential value at any point in the system is constant regardless of time.

On the other hand, the concept of the equilibrium state has been described in the discussion of thermodynamics and the definition of the equilibrium state is given as follows. A closed macroscopic system under consideration is said to be in equilibrium when the macroscopic physical quantity is, with great relative certainty, equal to its average value for any part of the system that is itself macroscopic. If the question is whether the system is in equilibrium or in non-equilibrium, one can use the criterion for non-equilibrium defined by

$$K_{\Delta P} = \frac{\Delta P}{P} \tag{1.72}$$

where P is the potential at an arbitrary point of the object, while ΔP is the difference (*or,* gradient) of potential values in the system. In particular, when the system is in equilibrium, it has the following characteristics under the condition $K \ll 1$:

(1) The difference (gradient) of potential values within the system is negligibly small. Therefore, the quantity that characterizes an equilibrium system has not the numerous values in a potential distribution, but only a single potential value.
(2) The quantities of state inside the system are at rest and there is no transfer or flow.

The definitions of the stationary state and the equilibrium state described above are very general and completely independent concepts. At first glance, the equilibrium state may be thought as a special case of the stationary state. Actually, there are some books on thermodynamics which treat the equilibrium state as a special case of the stationary state. However, since these two concepts are quite independent, there are four possible types for the classification of states of systems: stationary equilibrium, stationary non-equilibrium, non-stationary equilibrium, and non-stationary non-equilibrium states.

In a stationary equilibrium system, the state quantities are stationary and do not change with time. In this case, entropy is not generated. Stationary equilibrium systems are studied from the standpoint of the theory of equilibrium. In a stationary non-equilibrium system, the state quantity transfers but does not change with time. The inflow and the outflow of state quantities are equal at each instantaneous time. This case involves the generation of entropy. Stationary non-equilibrium systems are studied from the viewpoint of kinematics. In a non-stationary equilibrium system, state quantities change with time, but transport and entropy generation are ignored. This system is studied from the standpoint of statics. Finally, in a non-stationary non-equilibrium system, transport of state quantities and generation of entropy occur, and they change with time. This system is studied from the standpoint of dynamics.

1.8.10 Reversibility and Irreversibility

We have already mentioned some of the problems in relation to reversibility and irreversibility. The problem of this irreversibility is quite important when discussing chemical reactions, or more generally the changes of things. In this section, therefore, we will provide a summary discussion of the irreversibility problems that appear in the description system of reaction kinetics; quantum mechanics, classical mechanics, thermodynamics, and statistical physics.

Quantum Mechanics

First, let us look back at the irreversibility problem that has appeared in quantum mechanics. The theoretical system of quantum mechanics consists of two parts, a part related to the wave equation and a part related to the observation. As for the part related to the wave equation, the phenomena involved are only reversible and do not include irreversibility, as described above. This is because the wave equation is symmetrical about the reversal of time and if it is possible to change the state in a certain order, the state is also changeable in the completely reverse order. On the other hand, the part related to observation includes irreversibility, as described below.

We assume the eigenvalues of the physical quantity G are g_1, g_2, …, and the eigenfunctions for them are $\varphi_1, \varphi_2, \ldots$, and then φ is expanded in these eigenfunction groups:

$$\varphi = a_1\varphi_1 + a_2\varphi_2 + \cdots \tag{1.73}$$

where $G\varphi_K = g_K\varphi_K(K = 1, 2, \ldots)$. A group of pure states φ_1, φ_2, … appears when G is measured many times for the same state, which is represented by the wave function φ. The probability of appearance of each state is given by $|a_1|^2$, $|a_2|^2$, …, as described above. It should be noted that the wave function changes during the measurement. If g_K is obtained by measuring G, it means that the wave function φ shifts to φ_K. Moreover, it is a discontinuous, non-causal, irreversible change of the wave function from φ to φ_K caused by observation, and this cannot be controlled. All what we can do is to calculate the probability of transfer from φ to φ_K.

H-Function

Second, although not unique to quantum mechanics, the function defined as follows for the state before and after the observation changes in one direction. *That is*, it is assumed that the probability that the eigenvalues $g_1, g_2, \ldots$ are obtained by measuring a certain physical quantity G is w_1, w_2, …, and the corresponding eigenfunctions are represented by φ_1, φ_2, …. The state after the measurement is generally a mixed state, and the Boltzmann's H-function is defined for the mixed state:

$$H = \sum_i w_i \log w_i \tag{1.74}$$

Now, suppose that in the mixed state after the measurement of A, a quantity of B different from A (generally a quantity that cannot be commutative with A) is measured. Then, it can be proved that the value of the H-function in the mixed state resulting from the measurement of B is generally smaller than the value in the mixed state caused by the measurement of A. The H-function generally decreases by repeating the measurement in this manner (the H-function increases when a minus sign is added to H, so this can be formally regarded as entropy). From this, it can be said that the operation of measurement causes an irreversible change in the system. However, just keep in mind the following: The irreversibility of the H-function has its source in its statistical processing, and therefore this irreversibility is not inherent in quantum mechanical observations, but the same tendency can be seen in the classical case.

Classical Mechanics

Third, the question is whether irreversibility appears in classical mechanics or not. In conclusion, all the phenomena described by classical mechanics are reversible.

It is clear from the fact that the basic equations of classical mechanics are invariant with respect to time reversal, *that is*, change in the sign of time.

Thermodynamics

Finally, let us consider irreversibility in thermodynamics. The second law of thermodynamics, the law of increase of entropy, reflects the irreversibility of thermal phenomena. In the following, we will add some considerations to the nature of entropy in thermodynamics. Entropy is a quantity relevant for a system and must be locally confined to the internal region of the system. Entropy is the reduced calorific value obtained by dividing the amount of heat $d'Q_{\mathrm{rev}}$ that the system reversibly absorbs by the temperature T at that time, regardless of whether the actual reaction proceeds reversibly or irreversibly, and is defined as

$$\mathrm{d}S = \frac{\mathrm{d}'Q_{\mathrm{rev}}}{T} \tag{1.75}$$

The entropy defined in this way is a state quantity and does not depend on the pathway of change. The concept of entropy was then given a statistical-mechanical interpretation, defined as the number of microscopic states realizing a macroscopic state, and given a stochastic character. That is,

$$S = k\ln \Delta\Gamma = -k\overline{\ln w_n} = -k\sum_n w_n \ln w_n \tag{1.76}$$

Once a stochastic interpretation of entropy is given, it has been applied not only to thermodynamic phenomena but also to information theory. The H-function (Eq. 1.74) that appeared in quantum mechanics as the measurement problem described above is an example for applying this statistical-mechanical definition of entropy.

As described above, the second law of thermodynamics originated from the empirical cognition of the irreversibility of thermal phenomena and was formulated as the law of increase of entropy in isolated systems. This law was interpreted by statistical mechanics that the system always changes toward a more certain state. This second law of thermodynamics is used to know the direction of change in an isolated system. As for the problem to determine whether a change actually occurs or not, this law is used to formulate only necessary conditions but is useless to formulate sufficient conditions. Also, the definition of entropy ultimately assumes reversibility, i.e., equilibrium, which is crucial to the nature of thermodynamics, resulting in making this scientific system powerless for non-equilibrium processes. Finally, in the statistical-mechanical interpretation of entropy, it should be pointed out that the source of irreversibility lies not in dynamics but in statistics. In other words, irreversibility does not appear in individual dynamic phenomena, but rather comes from the phenomenon of statistical ensemble consisting of a huge number of particles.

Measure of Irreversibility

Then, let us concludes our discussion of the problem of irreversibility that appears in quantum mechanics, classical mechanics, thermodynamics, and statistical physics. Well, we will describe the discriminant or the measure of irreversibility. The discriminant of irreversibility is defined by

$$K_d = \frac{Q_d}{Q''} = \frac{\Delta P \Delta E}{P'' \Delta E} = \frac{\Delta P}{P''} \tag{1.77}$$

where Q_d is the amount of heat dissipated, Q'' is the work of inflow of state quantity, ΔE is the inflow of state quantity, $\Delta P \equiv P' - P''$, where P' and P'' are the potentials at the exit and entrance, respectively. In particular, $K_d \ll 1$ when the process is reversible.

Next, let us give a definition of the measure of the relative irreversibility of the conduction (in the system) and the transfer (to the surface of the system) of state quantities.

$$K_{d,0T} = \frac{K_{d'}}{K_{d''}} = \frac{\Delta P}{\delta P} \tag{1.78}$$

where a measure of irreversibility of conduction processes in a system is given by

$$K_{d'} = \frac{Q_{d'}}{Q''} = \frac{\Delta P \Delta E}{P'' \Delta E} = \frac{\Delta P}{P''} = \frac{\text{(dissipated heat due to conduction)}}{\text{(work of inflow of state quantity)}} \tag{1.79}$$

and a measure of the irreversibility of transfer processes on the surface of the system is given by

$$K_{d''} = \frac{Q_{d''}}{Q''} = \frac{\delta P \Delta E}{P'' \Delta E} = \frac{\delta P}{P''} = \frac{\text{(dissipated heat due to transfer)}}{\text{(work of inflow of state quantity)}} \tag{1.80}$$

In these definitions, if $K_{d,0T}$ is large, dissipated heat due to conduction is large compared to transfer. if $K_{d,0T}$ is small, dissipated heat due to conduction is small compared to transfer.

1.8.11 Continuity and Discontinuity

In quantum mechanics, physical quantities change discontinuously, as can be seen from the quantization of the eigenvalues of the Hamiltonian in a bound state. In the human world, on the other hand, such discontinuous changes are rarely encountered. Would it become a case that discontinuous changes in micro-system correspond to continuous changes in macro-system? In this section, we will consider the relationship between such continuity and discontinuity. We will show the first case:

the relationship between continuity and discontinuity. When the size of the space in which the quantum particles move becomes macroscopic, the intervals between discrete values of physical quantities become negligibly small and can be regarded as a continuous change.

1. **Discrete Energy Levels**

In the first place, we will pick up a simple but specific example to calculate discrete energy levels of a particle using quantum mechanics. We will consider the motion of a particle bound in the area by reflective walls that are boundaries separating the region of the constant potential energy. Let us suppose for simplicity that the potential energy $V(x) = 0$ at $-a < x < +a$ and $V(x) = +\infty$ at $|x| > a$ as shown in Fig. 1.6, and that completely rigid walls which any matters cannot pass through are set up at $x = \pm a$. In this case, the wave function must be zero at the positions of $x = \pm a$.

The wave equation for $|x| < a$ has such a simple form as

$$-\frac{\hbar^2}{2m}\frac{\mathrm{d}^2\psi}{\mathrm{d}x^2} = E\psi. \tag{1.81}$$

This has a general solution of

$$\psi(x) = A\sin\alpha x + B\cos\alpha x, \alpha = \sqrt{\frac{2mE}{\hbar^2}}. \tag{1.82}$$

Applying the boundary condition of $\psi = 0$ at $x = \pm a$,

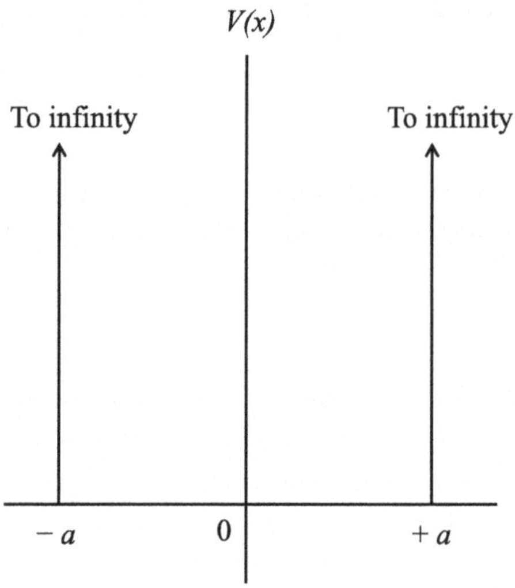

Fig. 1.6 One-dimensional square well-type potential

$$A \sin \alpha a + B \cos \alpha a = 0,$$
$$-A \sin \alpha a + B \cos \alpha a = 0 \tag{1.83}$$

From these, we obtain

$$A \sin \alpha a = 0 \text{ and } B \cos \alpha a = 0 \tag{1.84}$$

It is not appropriate that both A and B are zero, because the solution of $\psi = 0$ everywhere is physically meaningless. Furthermore, $\sin \alpha a$ and $\cos \alpha a$ cannot be simultaneously zero for a given value of αa or E. Therefore, the following two sets of solutions are possible:

The first set: $A = 0$ and $\cos \alpha a = 0$
The second set: $B = 0$ and $\sin \alpha a = 0$.

This leads to

$$\alpha a = n\pi/2. \tag{1.85}$$

Here, n is an odd number in the first set and an even number in the second set. Then, the two sets of solutions and their energy eigenvalues are

$$\psi(x) = B\cos\frac{n\pi x}{2a} \quad (n\text{:odd number}) \tag{1.86a}$$

$$\psi(x) = A\sin\frac{n\pi x}{2a} \quad (n\text{:even number}) \tag{1.86b}$$

In both cases energy is given by

$$E_n = \frac{\pi^2 \hbar^2 n^2}{8ma^2} (n = 1, 2, 3, \ldots) \tag{1.87}$$

It is clear that $n = 0$ gives a physically meaningless solution $\psi = 0$, and that a solution with a negative n is not independent of a solution with a positive n. It is not difficult to choose the constants A and B so as to normalize the eigenfunction in each case.

Thus, there is an infinite series of discrete energy levels corresponding to any positive integer quantum number n. There is exactly one eigenfunction for each level, and the number of nodes that the eigenfunction creates within the potential well is equal to $n - 1$. To be interesting, it is shown that the magnitude of the lowest energy level (i.e., the energy level of the ground state) satisfies the uncertainty relation (Eq. 1.11). In other words, if the round value of position uncertainty is a, the momentum uncertainty is at least about $\hbar/a$, and then this leads to a result that the lowest kinetic energy is about $\hbar^2/(ma^2)$.

Now let us return to the main subject. From the energy levels Eq. (1.87), the interval between the discrete energy values becomes nearly

$$\Delta E_n = \frac{\pi^2 \hbar^2}{8ma^2} \tag{1.88}$$

When this value of ΔE_n is calculated for electrons and we have

$$\Delta E_n = 2.35\,\text{eV} \quad \text{for } a = 2 \times 10^{-8}\,\text{cm},$$

and

$$\Delta E_n = 2.35 \times 10^{-16}\,\text{eV} \quad \text{for } a = 2\,\text{cm}.$$

Thus, the energy gap in an isolated atom or molecule is about 1–3 eV. In this case, the energy gap is not negligible, but for a metal piece of 1–2 cm, this energy gap is about 10^{-16} eV, almost negligible. Since squared a is found in the denominator of Eq. (1.88), $\Delta E_n \to 0$ rapidly with increasing a when $a \to \infty$. Thus, the energy actually can be considered to change continuously.

2. **Specific Heat in Solids**

The second interesting problem in the relationship between continuity and discontinuity arises, *for example*, in the theory of specific heat. The explanation of the theory of specific heat in classical statistical mechanics is based on the law of equipartition of energy. This law is as explained below [10, p. 3]. The total kinetic energy of a system (which is not the kinetic energy when the system moves as a whole, but the kinetic energy of the atoms and molecules that make up the system) is expressed by summing up the kinetic energy for each degree of freedom. If the kinetic energy belonging to each degree of freedom is proportional to the square of the velocity (or momentum) belonging to that degree of freedom, and if the dynamical system is immersed in a heat reservoir at temperature T, then the average value of the kinetic energy is given by $(1/2)kT$ for each degree of freedom.

In the thermal equilibrium state, such a circumstance will never happen that large amount of kinetic energy is distributed to a certain degree of freedom, and small amount of kinetic energy is distributed to the other degrees of freedom. This is the reason why this law is called "the law of equipartition of energy", and this law is the basis for the atomistic interpretation of temperature or heat. This law is derived as bellow. By assumption, the kinetic energy of the system has the form of

$$E_{kin} = \alpha_1 p_1^2 + \alpha_2 p_2^2 + \cdots + \alpha_s p_s^2 + \cdots + \alpha_f p_f^2 \tag{1.89}$$

where the general term $\langle \alpha_s p_s^2 \rangle$ is the kinetic energy of the s-th degree of freedom. The coefficient α_s may be a constant, but more generally, any non-negative function of $q_1, q_2, \ldots, q_f$. Provided that there is no interaction between molecules or atoms, the problem of determining the energy distribution of whole system can be replaced by the problem of determining the energy distribution of individual molecules and atoms. If the formula (1.42) is used to calculate the kinetic energy of each molecule or atom, the average value of $\langle \alpha_s p_s^2 \rangle$, kinetic energy of the s-th degree of freedom, is as follows:

$$\begin{aligned}
&\langle \alpha_s p_s^2 \rangle \\
&= \frac{1}{z} \iint \cdots \int_{-\infty}^{\infty} \alpha_s p_s^2 \exp\left\{ -\frac{1}{kT} E(q_1, q_2, \ldots, q_f, p_1, p_2, \ldots, p_f) \right\} \\
&\qquad \mathrm{d}q_1 \mathrm{d}q_2 \ldots \mathrm{d}q_f \mathrm{d}p_1 \mathrm{d}p_2 \ldots \mathrm{d}p_f \\
&= \frac{1}{z} \iint \cdots \int_{-\infty}^{\infty} \alpha_s p_s^2 \exp\left\{ -\frac{1}{kT} \left(\sum_{s=1}^{f} \alpha_s p_s^2 + V \right) \right\} \mathrm{d}q_1 \ldots \mathrm{d}q_f \mathrm{d}p_1 \ldots \mathrm{d}p_f
\end{aligned} \tag{1.90}$$

where V is the potential energy, which is a function of only the coordinate q, and z is a normalization factor given by

$$z = \iint \cdots \int_{-\infty}^{\infty} \exp\left\{ -\frac{1}{kT} E(q_1, q_2, \ldots, q_f, p_1, p_2, \ldots, p_f) \right\} \mathrm{d}q_1 \mathrm{d}q_2 \ldots \mathrm{d}q_f \mathrm{d}p_1 \mathrm{d}p_2 \ldots \mathrm{d}p_f \tag{1.91}$$

To obtain the mean value of $\langle \alpha_s p_s^2 \rangle$, let us calculate the following integral:

$$\int_{-\infty}^{\infty} \alpha_s P_s^2 \exp\left(-\frac{1}{kT} \alpha_s p_s^2 \right) \mathrm{d}p_s,$$

and we can get easily

$$\int_{-\infty}^{\infty} \alpha_s p_s^2 \exp\left(-\frac{1}{kT} \alpha_s p_s^2 \right) \mathrm{d}p_s = \frac{kT}{2} \int_{-\infty}^{\infty} \exp\left(-\frac{1}{kT} \alpha_s p_s^2 \right) \mathrm{d}p_s \tag{1.92}$$

Therefore, the mean value can be expressed as:

$$\langle \alpha_s p_s^2 \rangle = \frac{1}{z} \frac{kT}{2} \iint \cdots \int_{-\infty}^{\infty} \exp\left\{ -\frac{1}{kT} \left(\sum_{s=1}^{f} \alpha_s p_s^2 + V \right) \right\} \mathrm{d}q_1 \ldots \mathrm{d}q_f \mathrm{d}p_1 \ldots \mathrm{d}p_f \tag{1.93}$$

By using Eq. (1.91), it can be proved that

$$\langle \alpha_s p_s^2 \rangle = \frac{1}{2} kT \tag{1.94}$$

Table 1.1 Degrees of freedom for various systems

System	Degree of freedom			
	Translation	Rotation	Vibration	Total
Monoatomic molecule	3	0	0	3
Diatomic molecule	3	2	1	6
Triatomic molecule	3	3	3	9
⋮	⋮	⋮	⋮	⋮
n atomic molecule	3	3	$3n - 6$	$3n$

Since s can be anything from 1 to f, this equation proves the law of the equipartition of energy.

Now, we will summarize here about the degrees of freedom, which is important when applying the law of equipartition of energy. Let us consider a certain dynamic system here. If the positions in space of the particles, which make up this system and can be considered as mass points, are characterized by f coordinates, the system is said to have f degrees of freedom. First, consider the simplest system, *that is*, a mono-atomic molecule. Since this system is characterized by three coordinates in the x, y, and z directions, it has three degrees of freedom. If the system is a di-atomic molecule, there are two degrees of freedom of the zenith angle, θ, and the azimuth angle, φ, that determine the direction of the molecular axis, and three degrees of freedom of the coordinates of x, y, and z of the center of gravity of two atoms. Considering the vibration, i.e., the periodic change in the distance between the two atoms that compose the molecule, one degree of freedom is added. Thus, they totally give six degrees of freedom. In the same way, the degrees of freedom of the various systems are as shown in Table 1.1.

If the system has the degrees of freedom as shown in this table, as long as the system is in thermal equilibrium, energy of $kT/2$ is distributed equally to these degrees of freedom. The specific heat of various objects can be calculated immediately by using this law. *That is*, for a mono-atomic ideal gas, $(3/2)Nk = (3/2)R = 2.93$ cal/deg per mole. For a di-atomic ideal gas, since vibration is normally not considered, $C = (5/2)R = 4.96$ cal/deg. In a crystal, since only vibration is to be considered and 6 degrees of freedom of motion as a whole are excluded, the degree of freedom is $3N - 6$ per mole. In crystals, potential energy is also important in addition to kinetic energy, and the time average of both is equal. Each is $kT/2$, and therefore the specific heat $C = (3N - 6)(kT/2 + kT/2) \cong 3R = 5.85$ cal/deg. Specific heat of a crystal is empirically known as Dulong–Petit law. It closely matches the value of the total specific heat observed for many solids, including metals, from fairly high temperatures to near room temperature. However, at low temperatures the agreement is poor; the specific heat becomes clearly smaller and approaches zero at $T \to 0$.

Einstein proposed a simple model to explain that the lattice specific heat at low temperatures is less than the measured value of $3R$ per mole at high temperature. In the model, the thermal properties of lattice vibration of N atoms are replaced by the properties of $3N$ independent one-dimensional harmonic oscillators of equal frequency. Furthermore, the energy of the frequency is quantized according to the quantum theory proposed by Planck for blackbody radiation. According to Planck, the energy E_n of the oscillator has only the following value:

$$E_n = nh\nu \quad n = 0, 1, 2, \ldots \tag{1.95}$$

The mean energy is given by Eq. (1.35) as

$$\overline{E} = \frac{\sum_{n=0}^{\infty} E_n \exp\left(-\frac{E_n}{kT}\right)}{\sum_{n=0}^{\infty} \exp\left(-\frac{E_n}{kT}\right)} = \frac{\sum_{n=0}^{\infty} (nh\nu) \exp\left(-\frac{nh\nu}{kT}\right)}{\sum_{n=0}^{\infty} \exp\left(-\frac{nh\nu}{kT}\right)} = \frac{h\nu}{\exp\left(\frac{h\nu}{kT}\right) - 1} \tag{1.96}$$

At high temperatures ($kT \gg h\nu$), the numerators can be expanded as

$$\exp\left(\frac{h\nu}{kT}\right) - 1 = 1 + \frac{h\nu}{kT} + \cdots - 1 \cong \frac{h\nu}{kT} \tag{1.97}$$

Thus, $\overline{E} = kT$ is obtained. In other words, when the energy of the harmonic oscillator is quantized according to Planck, the value of the average energy at a high-temperature approaches the average energy, kT, in classical statistics.

At low temperatures ($kT \ll h\nu$), the situation is different.

$$\overline{E} = h\nu \exp\left(-\frac{h\nu}{kT}\right) \tag{1.98}$$

and the specific heat is given by

$$C_v \cong 3Nk\left(\frac{h\nu}{kT}\right)^2 \exp\left(-\frac{h\nu}{kT}\right) \tag{1.99}$$

Therefore, it approaches zero when $T \to 0$. This model is simple, but with a proper choice of frequency ν, the decrease in specific heat at low temperatures is fairly well represented. To compare the specific heats of various solids, it is convenient to use the characteristic temperature defined below (called the Einstein temperature):

$$h\nu = k\Theta_E \tag{1.100}$$

For many solids, Θ_E is between 100 and 300 K. When $\Theta_E = 300$ K, the characteristic frequency is about 5×10^{12} cps. This value is appropriate as the frequency of the vibration of the atoms in the solid and is close to the frequency of the acoustic vibration having a wavelength equal to the interatomic distance.

Thus, two treatments for the specific heat of solids have been presented, namely classical statistics and quantum statistics. Comparing these two treatments, you will notice the following interesting relationship as for the concepts of continuity and discontinuity: *That is*, if a molecule or an atom is considered as a microscopic oscillator and the classical statistics is used assuming a continuous change of the energy, an unnatural discontinuous result is obtained in macroscopic specific heat. On the other hand, assuming that the oscillator has only discrete values, it guarantees a rather continuous result for macroscopic specific heat of solids. *For example*, let us consider the cases of an N-atoms crystal, (A) where the distance between atoms is strictly constant and the atoms do not oscillate at all, or (B) where the atoms oscillate regardless of the magnitude of the frequency. The specific heats of the crystals of case (A) and (B) will be distinctly different.

We want to regard the limit where the atomic bonds are strong, and the frequency is gradually increased as the case in which the distance is constant. However, when classical statistics are applied to this oscillator, the specific heat is $3R$ for the case where it oscillates, while it is 0 for the case of non-oscillation. A clear discontinuous fault appears, and the limit-operation becomes meaningless. However, when Planck's quantum hypothesis is applied to this oscillator, the specific heat of the crystal becomes a continuous function of the frequency (Eq. 1.98), and when performing the limitation for approaching the infinitely large frequency, this agrees with the result obtained for the crystal of assuming non-oscillation dealt with classical statistics. The measured results also show that the continuous change in specific heat based on Planck's quantum hypothesis is correct. In order to avoid the occurrence of discontinuity in macroscopic phenomena, we must acknowledge the necessity of the assumption of discontinuity in microscopic phenomena. In other words, discontinuous changes in microscopic physical quantities lead to continuous changes in macroscopic physical quantities.

1.8.12 Classification of Physical Quantities

Up to the previous section, we have considered quantum mechanics, classical mechanics, thermodynamics, and statistical physics as description systems for reaction kinetics and dealt with various physical quantities in connection with these description systems. In this section, we will categorize these physical quantities from several viewpoints and summarize some of their characteristics.

1. **Classical and Quantum Quantities**

Here, the classical quantity means a physical quantity when relativistic and quantum effects are ignored, that is, a physical quantity dealt with non-relativistic and non-quantum mechanics. Therefore, the classical quantity has a definite instantaneous value and changes continuously. On the other hand, when the quantum quantities are conjugate to each other, they satisfy the uncertainty relation. Therefore, the quantum values are accompanied by quantum fluctuations, so that the instantaneous value is indefinite and changes discontinuously. As mentioned earlier, classical and quantum quantities can be linked by operations of the classical limit ($h \rightarrow 0$), by correspondence principle, and by wave function averaging. In such a case, the classical quantity and the quantum quantity are considered to have counterparts to each other, but sometimes they are not. As for spin and isospin in quantum mechanics, their counterparts cannot be found in classical physics and should be considered pure quantum quantities.

Quantum quantities are used exclusively for describing microscopic phenomena. Therefore, in that sense, they should be purely micro-quantities. In contrast, classical quantities are used to describe macroscopic phenomena in classical mechanics and are used to represent micro-quantities in classical statistical mechanics, especially in statistical thermodynamics. Thus, classical quantities are sometimes treated as macro-quantities and sometimes as micro-quantities.

Now, both classical and quantum quantities have a special relationship to thermodynamic quantities. *For example*, the concept of energy exists both in classical and quantum mechanics, and also in thermodynamics. But their characteristics are different. There is no special restriction on the energy in classical mechanics, *for example*, kinetic energy or potential energy. However, energy in quantum mechanics is defined only in a stationery state, and energy in thermodynamics is a state quantity defined in an equilibrium state and is a function of temperature and pressure. Thus, the characteristics of the same term "energy" vary depending on the theoretical system in which it is used.

2. **Thermodynamic and Non-thermodynamic Quantities**

Thermodynamic quantities are quantities with special characteristics as described below.

First, the thermodynamic quantity is indeed a typical macro-quantity. Some thermodynamic quantities as macro-quantities have their counterparts in micro-quantities such as quantum mechanics and classical mechanics, while others do not. As already mentioned, dynamic quantities of quantum mechanics or classical mechanics are linked to thermodynamic quantities by averaging operation using wave functions or Gibbs distribution. Energy, volume, pressure, velocity, etc. are typical examples, and as for these thermodynamic quantities, we can find their counterparts in quantum mechanics and classical mechanics. They are thermodynamic quantities which have dynamic properties. We have already pointed out that these dynamical quantities differ in their character depending on whether they are used in quantum mechanics, classical mechanics, or thermodynamics.

On the other hand, temperature and entropy cannot find their counterparts in quantum mechanics. In other words, temperature and entropy do not emerge from simply adding averaging operations to micro-quantities in quantum mechanics or classical mechanics. These are not, however, completely independent of the averaging operation. The entropy is given by Eq. (1.76).

$$S = k \ln \Delta \Gamma = -k\overline{\ln w_n} = -k \ln w_n\left(\overline{E}\right) \tag{1.101}$$

and is obviously related to the average. The temperature is defined in such way as Eq. (1.101), using the entropy thus defined and the energy E as an average value in Eq. (1.35).

$$T = \left(\frac{\partial E}{\partial S}\right)_P \tag{1.102}$$

Therefore, temperature is also indirectly based on the averaging operation. Other macro-quantities that seemingly unrelated to the average are also considered as those per unit volume, per unit area or per unit time, which are meant to be averaged for macroscopic volume, area, or time. Thus, we can say that thermodynamic macro-quantities are ultimately established based on averaging operation.

Second, thermodynamic quantities are defined based on the assumption of equilibrium. *For example*, definition of entropy, as we have already seen, ultimately assumes equilibrium states. By the way, if the system we are thinking of is not in equilibrium, it will change toward an equilibrium state. The time required for this system to transition to equilibrium is called the relaxation time, and this relaxation time increases as the size of a system increases. Therefore, if you talk about the entropy of a system that is not in equilibrium, the following must be done.

The problem is to examine a non-equilibrium system over a time Δt that is shorter than the relaxation time required for the system to reach equilibrium. To do this, the system must be divided into portions in such a way that the intrinsic relaxation time is smaller than Δt. With this partitioning, the subsystem can be regarded as being in some partial equilibrium described by a certain distribution function over the time Δt. Therefore, for these subsystems, entropy can be calculated according to the definition, and the entropy of the entire system is obtained as the sum of the entropies of the subsystems by using the additivity of entropy. The entropy of a system that is not in equilibrium can be obtained by dividing the system into parts, but the assumption of equilibrium in the definition of entropy, and thus the importance of the role of time, does not change at all. In order to determine the entropy of a non-equilibrium system, the system is divided such that the relaxation time is shorter than the given time Δt. However, when we consider the subsystem by dividing the system in question, this subsystem must itself be macroscopic. Thus, it is clear that for very short periods of time, entropy loses its physical meaning. In particular, we must not say anything like "instantaneous value" of entropy.

Third, thermodynamic quantities are state quantities, and then they are a function of temperature and pressure. Thermodynamic quantities are defined for equilibrium state as described above. The equilibrium state does not depend on the process in which it is realized. Therefore, the thermodynamic quantity does not depend on the path of the change or the rate of the change, but is a state quantity defined by temperature, pressure, and so on. Though they are called the same "energy", the energy in thermodynamics is a function of temperature and pressure defined only by the equilibrium state, which is different from the energy in quantum mechanics or in classical mechanics.

Fourth, as already clarified, the fluctuations in thermodynamic quantities involve quantum fluctuations as well as classical fluctuations. The system targeted by thermodynamics consists of an enormous number of particles, and the particles are not limited to classical particles or quantum particles and may be either. From the essence of physical phenomena, quantum fluctuations are extremely important, even if their magnitudes are smaller than classical fluctuations. This is because the equation of state for an object can be given by a quantum mechanical equation in some cases.

Fifth, thermodynamic quantities are considered to be completely continuous or almost continuous because the object is a macroscopic system.

The characteristics described above show how unique a thermodynamic quantity is when compared with a non-thermodynamic quantity such as the strength of an electromagnetic field.

3. **Microscopic and Macroscopic Quantities**

A typical example of a microscopic quantity is a quantum mechanical quantity as described above. On the other hand, the typical macroscopic quantities are thermodynamic quantities. Classical mechanical quantities are sometimes used as microscopic quantities and also as macroscopic quantities. If there are counterparts between the micro- and macroscopic quantities, they are connected with each other by averaging operation using wave functions or the Gibbs distribution. Purely macroscopic concepts, such as temperature or entropy, which have no direct micro-equivalents are also defined directly or indirectly by averaging operation. Non-thermodynamic and macroscopic concepts such as the strengths of electric field, magnetic field and so on are also defined based on the averaging over narrow space and short time. Such a macroscopic quantity can be said in short as something like an averaged grasp of the physical quantity. The source of this averaged cognition of macroscopic quantities is thought to come from the fact that we, the cognitive subjects, are the beings who can be distinguished by our human senses, and this situation may be inevitable for our human cognitive system.

References

1. S. Watanabe, *The History of Time* (Tokyo Tosho, Tokyo, 1975) (in Japanese)
2. T. Iwasaki, S. Miyahara, *Modern Natural Science and Material Dialectics* (Ohtsuki Shoten, 1972) (in Japanese)
3. L. Pauling, *The Nature of the Chemical Bond*, 3rd edn. (Cornell University Press, New York, 1960)
4. Л.Д. Ландау, Е.М. Лифшиц, *Механика*, ed. by T. Hiroshige, I. Mito (Tokyo Tosho, Tokyo, 1967). [English version] L.D. Landau, E.M. Lifshitz, *Mechanics*, 3rd edn. (Butterworth-Heinemann, 1982) (Japanese version
5. R. Carnap, *Philosophical Foundations of Physics: An Introduction to the Philosophy of Science* (Basic Books, 1966)
6. A. Messiah, *Quantum Mechanics* (Dover Publication, New York, 1995)
7. G.M. Barrow, *Physical Chemistry*, 2nd edn. (McGraw-Hill Book Company, 1966)

8. Л.Д. Ландау, Е.М. Лифшниц, *Статистйческая Физика*, ed. by A. Kobayashi (Iwanami Shoten, Tokyo, 1966) (English version) L.D. Landau, E.M. Lifshitz, *Statistical Physics*, 3rd edn. (Pergamon Press, 1980) (Japanese version)
9. Л.Д. Ландау, Е.М. Лифшниц, *Квантовая Механика*, translated into Japanese by K. Sasaki, S. Koumura (Tokyo-Tosho, Tokyo, 1971), (English version, L.D. Landau and E.M. Lifshitz *Quantum Mechanics*, 3nd edn. (Pergamon Press, 1981).
10. S. Tomonaga, *Quantum Mechanics*, 2nd edn. (Misuzu Shobo, Tokyo, 1974) (in Japanese)

Further Readings

F. Engels, *Dialektik der Natur* (Dietz Verlag, Berlin, 1952)
S. Sakata, *A New Wind for Science* [in Japanese] (Shin-nihon Publishing, Tokyo, 1966)
M. Bunge, *Causality—The Place of the Causal Principle in Modern Science* (Harvard University Press, 1959)
R.P. Feynman, *How were Physical Law Discovered?* This Japanese book included two papers, the Character of Physical Law and the Development of the Space-Time View of Quantum Electrodynamics, translated into Japanese by H. Ezawa (Diamond Company, Tokyo, 1968)
J. Monod, *Le Hasard et la Nécessité* (Alfed A. Knopf, Paris, 1970)
T. Dantuik, *Number, The Language of Science*, (Pearson Education, Inc., 2005)
R. E. Peierls, *The Law of Nature* (George Allen & Unwin, 1955)
H. Poincare, *La Science et l'Hypothèse*, translated into Japanese by I. Kono (Iwanami Shoten, Tokyo, 1978)
H. Poincare, *Science et Méthode*, translated into Japanese by Y. Yoshida (Iwanami Shoten, Tokyo, 1987)
T. Bastin, *Quantum Theory and Beyond* (Cambridge University Press, Cambridge, 1971)
K. Hushimi, M. Yanase, *What Is Time?* (Chuo-kouronsha, Tokyo, 1974) (in Japanese)
M. Taketani, *Contemporary Theoretical Issues* (Iwanami Shoten, Tokyo, 1968) (in Japanese)
K. Tomiyama, *Logic of Modern Physics* (Iwanami Shoten, Tokyo, 1970) (in Japanese)
Э.В. Шпольский, *Введеный в Атомную Физика,* ТОМ РЕРВЫЙ (Госурственное издательство, Москва 1963); E.V. Shpolsky, *Introduction to Atomic Physics*, vol. 2 (State Publishing House, Moscow 1963), by H. Tamaki et al. (Tokyo-Tosho, Tokyo, 1966) (Japanese version)
K. Amano, *History of Quantum Mechanics* (Chuo-kouronsha, Tokyo, 1974) (in Japanese)
J.V. Neumann, *Die Mathematische Grundlagen der Quantenmechanik* (Springer Verlag, Berlin, 1932)
M. Taketani, *Natural and Social Sciences* (Keiso Shobo, Tokyo, 1970) (in Japanese)
M. Taketani, *Introduction to Science* (Keiso Shobo, Tokyo, 1970) (in Japanese)
Messiah, *Quantum Mechanics* (Dover Publication, New York, 1995)
D. Bohm, *Quantum Theory* (Prentice-Hall Inc., Hoboken, 1951)
M. Taketani, *Various Issues Related to Dialectics*, Collected Works No. 1 (Keiso Shobo, Tokyo, 1970) (in Japanese)
M. Jammer, *The Conceptual Development of Quantum Mechanics* (McGraw-Hill Book Company, 1966)
G. Araki, *Quantum Mechanics* (Baifukan, Tokyo, 1961) (in Japanese)
D. Bohm, *Causality and Chance in Modern Physics* (D. Van Nostrand Company, 1957)
Д.И. Блохинцев, *Принципиальные Вопрсы Квантовой Механикц*, наука (Москва, 1963)
D.I. Blokhintsev, *The Principle of Quantum Mechanic*, ed. by T. Fukuyama (Sougo Tosho, Tokyo, 1974 (Japanese version)
А.И. Вейник, *Термодинамика Необратимых Процессов*, (Издательство « наука и технйка», Минск 1966), A.I. Vejnik, *Irreversible Process of Thermodynamics*, ed. by M. Senoo, (Sougou-kagaku publishing, Tokyo, 1969) (Japanese version)

Prigogine, R. Defay, *Chemical Thermodynamics* (Éditions Desoer, Liége, 1950)
M. Senoo, *Introduction to Thermodynamics of Irreversible Processes* (Tokyo Kagaku-Doujin, Tokyo, 1972) (in Japanese)
R.W. Gurney, *Introduction to Statistical Mechanics* (McGraw-Hill Book Company, 1949)
P.W. Bridgman, *The Nature of Thermodynamics* (Harvard University, 1941)

Chapter 2
Critique of the "Theory of Rate Processes"

Abstract In the first half of this chapter, several questions about unnatural assumptions in formulation of the absolute reaction kinetics are discussed. That is, (1) the validity of using a particle model behaving in a classical-mechanical manner under the potential determined by the time-independent Schrödinger equation, (2) the validity of assumption that the activated complex in a transition state specified by temperature is in equilibrium with the reactant although temperature should be defined only in equilibrium states, (3) question about the implicit basic premise that the reaction up the energy mountain occurs faster than the reaction down the mountain, and (4) questions about the convertibility of the potential energies related to the motion of electrons and kinetic energies related to the motion of atomic nuclei. Then, the conventional nucleation-growth theories (NGTs) to explain the changes such as phase transformation, solidification, and crystal growth are discussed. What corresponds to the assumption of "activated complex" is the assumption of "nucleus" in those changes, In NGTs, generation of hypothetical nucleus is assumed which is accompanied by the increased free energy until a certain size is reached. This assumption leads to the strange conclusion that nucleation does not proceed spontaneously. In addition, the concept of "time", which must be included in kinetics, is not included and NGTs do not take into account quantum mechanics.

Keywords Activated complex · Activation energy · Frequency factor · London equation · Coulomb energy · Exchange energy · Eigenvalue equation · Directionality of change

In Chap. 1, we firstly discussed general problems on the hierarchical and historical development of the world (nature) and its scientific cognition in order to establish a perspective for clarifying the fundamental problem of reaction kinetics as a current scientific cognition. The scope of consideration was then limited to the direct background of reaction kinetics; the four disciplines related to reaction kinetics, i.e., quantum mechanics, classical mechanics, thermodynamics and statistical physics, and focused on (1) hierarchy of the substances to be discussed, (2) irreversibility, and (3) the explicit expression of rate, that is, the time rate of change. This Chapter aims to elucidate the fundamental problems of current reaction kinetics, based on

S. Yamamoto, *Reaction Kinetics Based on Time-Energy Uncertainty Principle*,
https://doi.org/10.1007/978-981-19-9673-3_2

these achievements in Chap. 1. At that time, rather than vaguely taking up the current reaction kinetics, Eyring's "The Theory of Rate Processes", which is famous as a typical theory of the current reaction kinetics and familiar as "absolute reaction kinetics", is selected and the problems involved in it will be analyzed. Later let us add some consideration of the "nucleation theory" which is closely related to this absolute reaction kinetics. The fundamental problems of the current kinetics clarified in this chapter are the current issues in the reaction kinetics that we should solve. A new attempt to solve these fundamental problems will be developed in Chap. 3.

2.1 Critique of the Absolute Reaction Kinetics

In the following sections, the problems in Eyring's absolute reaction kinetics will be specifically discussed.

2.1.1 Application Fields of the Absolute Reaction Kinetics cf. [1, pp. 1–2]

S. Arrhenius studied the effect of temperature on the conversion rate of sucrose and proposed the following equation in 1889:

$$\kappa = A\exp\left[-\left(\frac{E}{RT}\right)\right] \tag{2.1}$$

Here, κ is the specific reaction rate, E is the activation energy (difference in heat content between active molecules and inactive molecules), and A is the frequency factor. Arrhenius thought that there are inert and active molecules in the reactants, and that only the latter participates in the reaction. It is now generally accepted that this type of relational expression represents the temperature dependence of almost all chemical reactions and the specific rates of certain physical processes. If the range of temperature is not wide, the quantities A and E can be considered as constant.

Eyring's absolute reaction kinetics aims to theoretically calculate A and E in Eq. (2.1), thereby predicting the rates of chemical reactions by calculation alone. Therefore, if this theory is successful, it would be possible to calculate the reaction rates for reactions caused by the temperature elevation, that is, reactions proceeded by so-called thermal activation. However, we know that chemical reactions are caused not only by thermal activation but also by action of light, electrical action, magnetic action, and also mechanical action. In fact, for example, Eq. (2.1) is not adopted as the specific rate constant in the initial process of a photochemical reaction. In addition, Eq. (2.1) cannot be applied to chemical reactions that proceed in a non-equilibrium state, such as reactions in plasma. Even if Eyring's absolute reaction kinetics is successful, it is necessary to confirm that this theory cannot deal with the rates of all chemical reactions. Moreover, even in the case of reactions caused by thermal activation, the strict application of this theory can hardly be said to succeed not only for actual complicated chemical reactions but also for the simplest chemical reactions. Let us consider the primary reasons step by step in the following.

2.1.2 Assumption of Definiteness in Energy cf. [1, pp. 2–5, pp. 62–84, pp. 91–93]

In Eyring's absolute reaction kinetics, basically the method of calculating the activation energy E in Eq. (2.1) is based on the following equation proposed by F. London in 1928 (so-called London equation).

$$E = A + B + C - \left\{ \frac{1}{2}\left[(\alpha - \beta)^2 + (\beta - \gamma)^2 + (\gamma - \alpha)^2\right] \right\}^{1/2} \tag{2.2}$$

Here, E is the potential energy of a system of three atoms X, Y and Z, each having an uncoupled s-electron. A, B, and C are the energies of Coulomb interaction of electron pairs with the atoms X and Y, Y and Z, and Z and X, respectively, and correspondingly α, β, and γ are resonance or exchange energy that comes from the quantum mechanical consequences of inability to regard an electron as localized with respect to the nucleus that was associated in the separated atom. The values of A, B, and C and of α, β, and γ are determined depending on the interatomic distances, and therefore the potential energy surface should be obtained by Eq. (2.2), giving the variation in potential energy for all possible interatomic distances.

By the way, as for the method for actually obtaining the Coulomb energies A, B, and C and the exchange energies α, β, and γ, it is extremely difficult to solve the integrals that give these energies separately, even in the simplest case of two hydrogen atoms. Therefore, in the absolute reaction kinetics, it is obtained by the method known as a semi-empirical method. For example, the binding energy of a pair of atoms X and Y is given by $A + \alpha$, and the variation with distance is determined by Morse equation based on the spectroscopic measurement data. A is assumed to be a certain fraction (generally referred to as 10 ~ 20%) of the whole $A + \alpha$, and the individual values of A and α are obtained. Similarly, if the necessary spectroscopic data for atomic pairs YZ and ZX are available, B and β, and C and γ can be determined. Thus, Eq. (2.2) gives all the quantities to yield the potential energy surface. Using the potential energy surface obtained in this way, for example, the reaction between three atoms X, Y and Z is visualized as

$$\mathrm{X} + \mathrm{YZ} \rightarrow \mathrm{XY} + \mathrm{Z} \tag{2.3}$$

As the reactant X approaches YZ along the most favorable reaction pathway, the potential energy of the system initially increases slowly, then increases more rapidly to a maximum, and then decreases as the products XY + Z are formed. The highest point on this pathway is considered to give the position of an activated state, and the energy difference between this point and the level representing the unreacted substance is considered as equivalent to the activation energy. In this way the activation energy is calculated from the potential energy surface of the system under consideration.

This semi-empirical method has been criticized for the following reasons (the reader should be careful for the fact that the terminology of "semi-empirical method" is also used in molecular orbital calculation with a slightly meaning different from the original Eyring's usage). The first is the assumption that the total binding energy is divided into the Coulomb energy and the exchange energy at a certain ratio. Second, the following approximations were made in solving the eigenvalue equations of quantum mechanics, Eq. (2.4), to derive the London equation under several assumptions.

$$H\psi = E\psi \tag{2.4}$$

That is,

(1) Since the mass of a nucleus is larger than that of an electron, it is supposed that the nucleus is at rest, and therefore the kinetic energy of the nucleus is omitted.
(2) The electron orbital is not managed to accurately treat the mutual repulsion between electrons.
(3) Chemical reaction is regarded as an adiabatic process, that is, the process is assumed not to involve the transition of electrons and therefore the electronic state can be represented by a single function which can be applied throughout all processes of chemical reaction.
(4) The multiple exchange integration is ignored.
(5) The interaction between orbit and spin is ignored.

These are mainly approximation problems for solving Eq. (2.4), not belonging to problems for principles. It would be better if Eq. (2.4) could be solved without using such approximations. But even with such an approximated solution, it is sufficiently worthy one, provided the approximation is satisfactory.

The more essential and principal problem lies in the following points. It is the very idea of trying to describe the reaction process by the eigenvalue Eq. (2.4). If the reaction process can be described in principle by solving the eigenvalue equations, the problem is only how to solve the eigenvalue equations by a good approximation method. However, if the reaction process cannot be described by solving the eigenvalue equation, no matter how good the approximate solution of the eigenvalue equation is developed, it becomes meaningless even if an exact solution is obtained instead of an approximate one.

Then, is it possible to describe the reaction process by solving the eigenvalue equation? What does it mean at all to use the eigenvalue equation, and what can be described using the eigenvalue equation? Let us take a closer look at these points. As discussed in the logic of quantum mechanics in Sect. 1.8.1, the eigenvalue Eq. (2.4) is an equation for describing the stationary state of a system, not for the non-stationary state. The atomic reaction process as expressed by Eq. (2.3) is non-stationary state. Therefore, it is impossible in principle to describe the reaction process of atoms with the eigenvalue equation.

The absolute reaction kinetics analyzes the reaction process of atoms and explains the activation energy using the potential energy surface. The potential energy surface is made based on London Eq. (2.2), and the London equation is obtained by solving the eigenvalue equation that describes the stationary state. Thus, the potential energy surface should be primarily applied to the stationary state. In the absolute reaction kinetics, the potential energy surface that should be applied to the stationary state is applied to the non-stationary state or the transition state of atoms. Here lies the crucial difficulty of the theory. This difficulty is the same as that laid in the current reaction kinetics as long as the current reaction kinetics is based on the viewpoint of the absolute reaction kinetics. From the above, it was found that the description of the reaction process using the eigenvalue equation of atoms has a fundamental difficulty. We further consider this problem from another perspective.

As will be described in detail later in Sect. 4.1.2, the eigenvalue Eq. (2.4) describes the state of a system in which there is no quantum mechanical fluctuation of energy E. By solving the eigenvalue Eq. (2.4), a definite value without fluctuations as an eigenvalue is actually obtained for the energy of the system. Contrarily, we should note that as discussed earlier, the eigenvalue equation is an equation that describes the stationary state. Taking into this consideration into account, we can conclude that the system in the stationary state is described by the eigenvalue equation and the energy takes a definite value.

As already mentioned, non-stationary states cannot be described by the eigenvalue equation. Then, will the energy of system have a definite value in a non-stationary state? As will be described later in Sect. 3.1.3, the system does not take one definite value of energy in non-stationary states or transition states and is accompanied by quantum mechanical fluctuations. And as described in Sect. 4.1.3, non-stationary state or transition state is described as superposition of different energy states.

Thus, we may conclude from the above conventional discussion based on the activation energy in terms of potential energy surface is eventually originated from such assumption that the reaction process of atoms is describable using the eigenvalue equation and that the energy has a definite value without uncertainty. It is further concluded that the correct way of formulation is to describe the reaction process of atoms as a superposition of various energy states and to deal with energy as a quantity accompanied with quantum mechanical fluctuations. A novel formulation of the atomic reaction process based on this new idea will be given in Chaps. 3 and 4.

2.1.3 *Fundamental Laws Predicting Directionality of Irreversible Change cf. [1, pp. 185–187]*

In the absolute reaction kinetics, the reaction rate is calculated in the way of following scheme. First, we consider the processes involved in the reactants A, B, etc., and the activated complex $M^{\ddagger}$ is accordingly assumed to be formed in the reaction.

$$A + B + \cdots\cdots \rightarrow M^{\ddagger} \rightarrow \text{Product} \tag{2.5}$$

The reaction rate is calculated by using the concentration (number of molecules per unit volume) of the activated complexes at the top of the potential barrier multiplied by the frequency of passing the barrier. Taking $C'_{\ddagger}$ as the number of activated complexes in unit volume lying in the length δ, representing the activated state around the top of the barrier and taking $\overline{v}$ as the mean velocity crossing the top of the barrier, then $\overline{v}/\delta$ is the frequency at which the activated complexes pass over the barrier. Thus, the reaction rate is expressed as

$$\text{Reaction rate} = C'_{\ddagger}\frac{\overline{v}}{\delta} \tag{2.6}$$

Eyring calculated the mean velocity $\overline{v}$ of the activated complex passing over the top of the barrier by the method of classical mechanics as

$$\overline{v} = \frac{\int_0^{\infty}\exp\left[-\frac{1}{2}m^*v^2/kT\right]v\mathrm{d}v}{\int_{-\infty}^{\infty}\exp\left[-\frac{1}{2}m^*v^2/kT\right]\mathrm{d}v} = \sqrt{\frac{kT}{2\pi m^*}} \tag{2.7}$$

where the limits of integration in the denominator are taken from $-\infty$ to $+\infty$ because the situation that the activated complexes move in both positive and negative directions is taken into consideration. The mass m^* is considered as the effective mass of the activated complex.

Now the problem is concerned with the way to find such a velocity $\overline{v}$. If the reaction rate is calculated as in Eq. (2.6), the velocity $\overline{v}$ must be the difference between the velocities of the activated complexes moving in the positive and negative directions, that is, the net velocity. Is the calculation of Eq. (2.7) supposed to provide with the result of such a net velocity in the positive direction of reaction? Eq. (2.7) only finds the mean velocity of one particle when the system in equilibrium under temperature T contains a huge number of particles. The value of $\overline{v}$ in Eq. (2.7) is the mean velocity when the activated complex consisting of multiple particles is regarded as one particle, and m^* is the effective mass when the multiparticle system is regarded as one particle system.

As already clarified in the discussion in Sect. 1.8.2, the equation of motion in classical mechanics is invariant with respect to time inversion, that is, changing the sign of time. This symmetry in two directions of time means that the motion described by classical mechanics is essentially reversible. Therefore, if the activated complex moves in the positive direction at the mean velocity $\overline{v}$ calculated by Eq. (2.7), it can move in the negative direction at the same velocity $\overline{v}$. In fact, according to the properties of the potential surface elucidated by Eyring, the sum of potential and kinetic energies of the particles moving on the potential surface is preserved.

Suppose a case that a particle initially staying at the point P_1 of the potential surface shown in Fig. 2.1 begins to move to the right direction on the potential surface and passes the peak P_2. When the object reaches the point P_3 with the same potential energy as P_1, it turns to the left and passes through P_2 at the same speed when it passed to the right. In a general manner, if the movement to the right with velocity $\overline{v}$ is possible, the movement to the left with the same velocity $\overline{v}$ is also possible. Thus, the net reaction rate obtained from the deduction of positive and negative directions becomes zero. If the reaction rate is obtained by Eq. (2.6), its net velocity $\overline{v}$ should be estimated using a fundamental law that gives some indication for the direction of movement. However, in the calculation of Eq. (2.7), such law expressing the irreversibility of change is not used at all. We put stress on the fact that the $\overline{v}$ value in Eq. (2.7) is not the net velocity in the positive direction to be obtained.

Therefore, the absolute reaction kinetics built on such calculations has no basis for its validity. The above consideration shows that the difficulty of the absolute reaction kinetics in calculating the reaction rate is the lack of the fundamental law that predicts the irreversibility of change.

Next point we should consider is how we solve the problems built-in the absolute reaction kinetics. To overcome the difficulties, we need to find a law correctly dealing with the directionality of change, that is, a law expressing that a reaction irreversibly changes in one direction. Turning back to the analysis of the description system

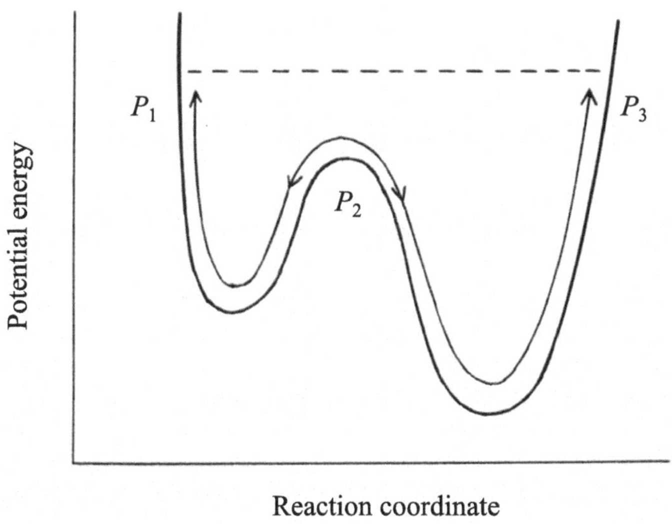

Fig. 2.1 Reversible motion of particles on potential energy surface

of reaction kinetics in Chap. 1, we can confirm that discontinuous, non-causal and irreversible changes of wave functions at any observations in quantum mechanics, unidirectional changes of Boltzmann's H-function, and the law of increase of entropy in thermodynamics are related to irreversibility of system.

However, these theories or laws do not include the concept of time required for constructing the reaction kinetics, accordingly they do not serve as the fundamental law we are looking for. After all, we finally reach the uncertainty relation of time and energy $\Delta t \cdot \Delta E \cong \hbar$ as such a fundamental law. A new reaction kinetics based on this time–energy uncertainty relation will be developed in Chap. 3.

2.1.4 Assumption of Equilibrium in Transition State cf. [1, pp. 13–14, pp. 100–107, p. 185]

In the absolute reaction kinetics, the equilibrium assumptions are used predominantly in the description of the formulation of reaction rates. The first usage appears in the formulation of the specific reaction rate k. In this formulation, it is assumed that the reactant is always in equilibrium state with the activated complex, which decomposes at a certain rate. It is also assumed that the reaction gives no appreciable influence on the equilibrium concentration of the activated complex. According to these fundamental assumptions in the absolute reaction kinetics, the specific reaction rate, k, is given by Eq. (2.9), where $K^{\ddagger}$ is the equilibrium constant of the reaction (2.8) between reactants and the activated complex.

$$\mathrm{A} + \mathrm{B} + \cdots \rightleftarrows \mathrm{M}^{\ddagger} \tag{2.8}$$

$$\kappa = \frac{kT}{h} K^{\ddagger} = \frac{kT}{h} \frac{C_{\ddagger}}{C_A \cdot C_B \cdot \cdots} \tag{2.9}$$

Using the well-known thermodynamic relations, the equilibrium constant $K^{\ddagger}$ can be rewritten as

$$\kappa = \frac{kT}{h} \exp\left[-\left(\frac{\Delta F^{\ddagger}}{RT}\right)\right] = \frac{kT}{h} \exp\left[-\left(\frac{\Delta H^{\ddagger}}{RT}\right)\right] \cdot \exp\left[\left(\frac{\Delta S^{\ddagger}}{R}\right)\right] \tag{2.10}$$

where $\Delta F^{\ddagger}$, $\Delta H^{\ddagger}$, and $\Delta S^{\ddagger}$ are changes in standard free energy, standard enthalpy, and standard entropy, respectively, at the formation of activated complexes from the reactants. Equation (2.10), which is a result of the formulation of the specific reaction rate κ, based on the assumption of equilibrium between reactants and activated complexes, is considered to correspond to the Arrhenius Eq. (2.1). Therefore, $\Delta F^{\ddagger}$ and $\Delta H^{\ddagger}$ in the Eq. (2.10) are supposed to play a role of the activation energy E in the Eq. (2.1). However, are there any crucial problems in the abovementioned fundamental premise and the formulation of specific reaction rates, Eq. (2.10), in the absolute reaction kinetics? Actually, there exist some crucial muddles, as described below.

1. **Problems in assumed rate-determining step in the absolute reaction kinetics**

Based on the fundamental premise in the absolute reaction kinetics, the whole process of chemical reaction from reactants to products comprises two processes of elementary reactions: (1) the reaction of forming activated complexes from reactants and (2) the reaction of decomposing the activated complexes into products. It is also assumed that these two elementary reactions proceed separately but in a serial relationship. Furthermore, an equilibrium is assumed to always hold between the reactants and the activated complexes, as represented by the Eq. (2.8). What does this mean? The situation that the reactants and the activated complex are always in equilibrium, namely, no matter at what rate the activated complexes decompose to products, means that the rate of reaction (1) (which is same as the rate of opposite reaction because of the equilibrium assumption) is larger than the rate of reaction (2). In other words, as long as this fundamental premise is satisfied, it means that the rate of overall reaction is controlled by the process of reaction (2).

A closer consideration turns out that the abovementioned fundamental premise is quite strange. Because the activated complex is in a high energy state, the reactants have to climb the mountain of energy to become the activated complex. On the other hand, the process of decomposition of the activated complex into a product is a process in which the energy decreases from the peak to the valley of the potential surface. Considering this pattern of energy variation through the reaction process, we naturally expect that the reaction (1) which corresponds to the formation of the activated complex from the reactants is slower than the reaction (2) which corresponds to the decomposition of the activated complex into products. This is exactly the opposite of what the fundamental premise implies. The fundamental premise that reactions of climbing up a hill of energy proceed faster than reactions of going down an energy slope is quite absurd. Such a scientifically strange assumption of the fundamental premise without justified reasons would distort the basis of the absolute reaction kinetics. In fact, the distortion of the theory has already appeared in the Eq. (2.10). The specific reaction rate, k, given by the Eq. (2.10), determines the overall reaction rate. However, the quantities of $\Delta F^{\ddagger}$, $\Delta H^{\ddagger}$, $\Delta S^{\ddagger}$, etc. that determine the value of k are for the reaction (1), and not for the reaction (2) which is slower than the reaction (1) and has a serial relationship with the reaction (2). This means that the overall reaction rate is controlled by a non-rate-determining reaction process. Such a contradiction is a direct consequence from the unnatural fundamental premise.

2. **Problems of equilibrium assumption between reactants and the activated complex**

Let us move to examine the physical contents of the above fundamental premise from a slightly different angle. The change in free energy associated with a chemical reaction is given by the well-known equation:

$$\Delta F = \Delta F^{\circ} + RT \ln \frac{a_{\ddagger}}{a_A a_B \cdots} \tag{2.11}$$

where ΔF° is the standard free energy and a_i is the activity of the i-th substance. The above fundamental premise assumes that the reactants and the activated complex are in equilibrium. Therefore, there is no change in free energy, $\Delta F = 0$, for the reaction (1). In contrast, the reaction (2) is out of equilibrium and then $\Delta F < 0$, i.e., a drop in free energy proceeds. If these results are illustrated in a commonly used expression, as shown in Fig. 2.2a, we can see no peak in free energy on the pathway from the reactant to the product (Strictly speaking, it is not correct to treat each instantaneous state of non-equilibrium reaction in this way, as will be discussed later). Any change that is thermodynamically possible in nature is accompanied by a decrease and without any increase in free energy. Therefore, no feature like a hill appears in the free energy profile.

The fundamental premise in Eyring's theory of absolute reaction rates is then probably introduced to derive such an equation of the form like Eq. (2.1) thermodynamically, by making an assumption of equilibrium between reactants and activated complexes, which resulted in an up-down shape in the standard free energy surface.

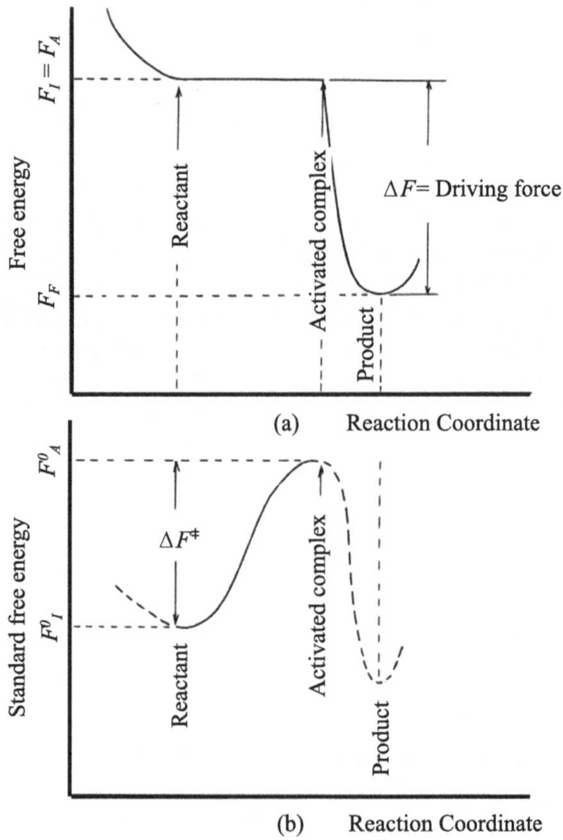

Fig. 2.2 Variation in free energy along reaction coordinate

Can we expect that a peak exists on the pathway from reactant to product, replacing a standard free energy instead of a free energy? If there would exist a peak of the change in standard free energy from reactants to products, it may be expressed as Fig. 2.2b. However, such a standard free energy curve has a direct effect for the reaction only in the case that an equilibrium exists not only between reactants and activated complexes, but also between activated complexes and products. That is, the standard free energy determines the equilibrium ratio of activities when the whole system is in an equilibrium state and the reaction does not proceed. If the system is not in equilibrium and the reaction proceeds, the concentration distribution could not be determined with the standard free energy. Under the fundamental premise in the absolute reaction kinetics, there is no equilibrium between activated complexes and products, and the reaction from activated complexes toward products proceeds. This reaction rate is equal to the rate of the formation of activated complexes from reactants minus the rate of its reverse reaction. Therefore, strictly speaking, there exists no equilibrium between reactants and activated complexes as long as the reaction is taking place as a whole.

The above consideration revealed that the formulation of the specific reaction rate expressed in Eq. (2.10) is usable only in a strict case of a complete equilibrium state or almost in equilibrium state where the reaction does not disturb the equilibrium between reactants and activated complexes. Thus, the Eq. (2.10) does not hold in reactions for which any equilibrium assumption should be prohibited. Such irrational fundamental premise in the absolute reaction kinetics brings about a significant irrationality in the free energy curve, the standard free energy curve, and also in the position corresponding to the activated complex on the potential surface. In other words, two (the classical and quantum?) worlds must be used properly. The right and left sides of the energetic hill are originally of the same nature and continuous, and therefore they are ought to be treated as the homogeneous world. However, due to the irrational fundamental premise, the world is split into two separate parts at this position (top of hill), and the parts are described by heterogeneous and discontinuous rules.

3. **Problems in interpretation of activation energy as thermodynamic quantities**

The problem which we should point out is that the parameters, $\Delta F^{\ddagger}$, $\Delta H^{\ddagger}$, and $\Delta S^{\ddagger}$ in Eq. (2.10) are assigned as the changes in free energy, enthalpy, and entropy for reactants and activated complexes in the standard state, respectively. Accordingly, the standard state itself of reactants or activated complexes can be arbitrarily selected for the convenience of calculation. Therefore, the values of $\Delta F^{\ddagger}$, $\Delta H^{\ddagger}$, and $\Delta S^{\ddagger}$ are arbitrarily chosen, depending on how to select the standard state. On the other hand, the activation energy we know empirically does not depend on the choice of the standard state of reactants or activated complexes. Therefore, the nature of the activation energy in the Eq. (2.1) and that of $\Delta F^{\ddagger}$, $\Delta H^{\ddagger}$, or $\Delta S^{\ddagger}$ in Eq. (2.10) are different from each other. In addition, what causes us to feel inconsistency between the activation energy of Eq. (2.1) and thermodynamic quantities such as $\Delta F^{\ddagger}$, $\Delta H^{\ddagger}$, and $\Delta S^{\ddagger}$ in Eq. (2.10) is related to the nature of the activation energy E as a physical quantity.

Thermodynamic quantities mentioned in Sect. 1.8.12 are (a) macroscopic quantities, and (b) state quantities defined for equilibrium states which are therefore functions of temperature and pressure. As for (a), thermodynamic energy seems not to be suitable for representing the activation energy, since such microscopic phenomena is related with thermal activation of individual atoms or molecules. As for (b), the activated state is unreasonably regarded as an equilibrium state. The Arrhenius activation energy is independent of temperature unless the temperature range is very wide, whereas the thermodynamic energy is a function of temperature and pressure.

As described above, the fundamental premise in the absolute reaction kinetics and the formulation of the specific reaction rate expressed by Eq. (2.10) based on this premise are involved with many crucial difficulties. Are there any good solutions to derive Arrhenius Eq. (2.1) without using such a problematic premise? Fortunately, we may expect an optimistic answer on this point and will derive Arrhenius formula later in Sect. 4.2 based on the uncertainty relation $\Delta t \cdot \Delta E \cong \hbar$. In addition, a more detailed examination about the problems in the conventional way of thinking on the thermal activation and the Arrhenius equation will be carried out.

In the absolute reaction kinetics, the second equilibrium assumption used for formulating the reaction rate relates to the temperature equilibrium in the transition state. A typical example is found in the calculations of the reaction rate already described in Sect. 2.1.3. In calculating the mean velocity $\overline{v}$ of the activated complex crossing over a top of the barrier in Eq. (2.7), it is supposed that the activated complex is at the temperature T. However, is it possible to define temperature for the activated complex in the transition state? Eventually there exists a problem at this point, as described below. Eyring first described that the sum of the potential and kinetic energies is preserved in the motion of a particle on the potential surface. The law of conservation of energy and the assumption of temperature equilibrium in the transition state are not compatible with each other as the explanation of the thermal activation described below.

We shall consider a potential surface as shown in Fig. 2.3. It is assumed that the particle stands still at the potential valley P_2. If the particle gains kinetic energy for some reason, it will be possible to imagine the situation that the particle climbs along the potential slope to a height equal to its kinetic energy from the viewpoint of classical mechanics. The thermal activation is supposed to be such process that this kinetic energy is given by the thermal energy and thereby the passing over the potential peak may become possible. In this case, a particle staying at the potential valley P_2 will be able to move to the potential valley P_4 beyond the peak P_3 only when the kinetic energy is given by the thermal energy becomes higher than the height of the peak. Suppose that the energy of thermal motion obtained by the particle that was initially stand still at the potential valley P_2 is exactly equal to the height of the potential peak. Then, when this particle moves on the potential surface and reaches the peak P_3 of the potential mountain, the kinetic energy of particle becomes zero, that is, the kinetic energy in the transition state becomes zero. This corresponds to the situation that the temperature of the particle in a transition state is equivalent to 0 K, when converted to temperature according to classical statistical mechanics.

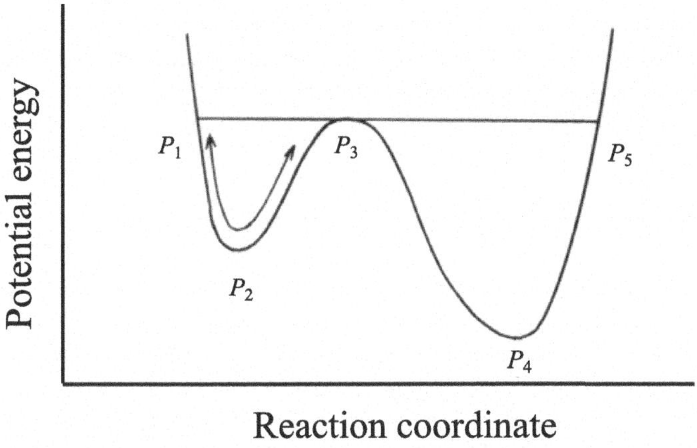

Fig. 2.3 Thermally activated motion of particles on potential energy surface

If we consider that the reactants overcome the energy barrier of transition state the state of energy maximum along the reaction path, by thermal activation during the progress of the chemical reaction, it is necessary to suppose that the temperature of T under which the particle was at the potential valley P_2 becomes 0 K in the transition state. Of course, particles at any point on the potential surface, such as P_1, P_2, P_3, P_4 etc., can also be supposed to be at T. Taking this absurd situation into account, we must conclude that the particle wherever being on the potential surface always has a kinetic energy higher than the potential energy by the amount equivalent to thermal energy, and the peak of the potential does not become an energetical obstacle preventing for the particle from transition. Thus, if we adopt the Eyring's basic scheme to explain the thermally activated process by the transition of particles over the transition state (or, the saddle point in potential surface), which is the state corresponding to the highest potential energy along the reaction coordinate, we have to consider that temperature can be defined only in the equilibrium state where the particle is in the bottom of the potential valley and that the temperature cannot be defined in the transition state.

As is suggested in the above discussion, in the particle motion on the potential surface, the law of conservation of energy, which means that the sum of the potential energy and the kinetic energy is preserved, and the assumption of temperature equilibrium in the transition state are incompatible. Thus, we need recognize that transition is essentially a non-equilibrium phenomenon.

In the case of non-equilibrium transition, the temperature cannot be defined, as already revealed in Sect. 1.8.3. As already mentioned, the calculation of $\overline{v}$, Eq. (2.7), for determining the reaction rate is unacceptable, but the same conclusion is obtained from the consideration in this Section. That is, it is clear that such a calculation as the Eq. (2.7) under the assumption of temperature equilibrium in the transition state cannot be no more accepted after the above discussion. Then, next question may emerge whether we can explain the process of thermal activation without assuming the temperature equilibrium in the transition state or not. In Sect. 4.2, we will show that thermal activation can be explained by assuming the temperature equilibrium only in the prereaction state, but not in the transition state.

2.1.5 *Problems Related to Hierarchy in Cognition*

In Chap. 1, we stated that the world (nature) has been interpreted with concepts in various levels of hierarchy, and that the theoretical systems have been established to describe the unique movement in each level of hierarchy belonging to subjects of our cognition. The phenomena in the chemical reaction, i.e., chemical changes, reaction rates, mass transfer, and so on, are belonged to the hierarchy of atoms and electrons. Therefore, chemical reaction kinetics is an issue how we humans recognize the reactions in the hierarchical level of atoms and electrons. As is revealed from the discussion of 1.8.1 through 1.8.4, quantum mechanics is a theoretical system that mainly describes the microscopic phenomena in the hierarchy of atoms and electrons, and classical mechanics and thermodynamics are for describing the macroscopic matters which belong to the hierarchy of our human world. The statistical physics is the way to describe how these two systems are linked. Considering the ways to approach the absolute reaction kinetics from the viewpoint of such concepts of hierarchy, we find the following problems.

1. **Uncritical mixing of the term "energy" in quantum mechanics, thermodynamic, and statistical physics**

Looking at the treatment of activation energy in the absolute reaction kinetics, the approach can be roughly categorized in three manners: The first type of the category is application of the potential energy surface [1, pp. 2–4, pp. 85–91], the second one is based on thermodynamics, and the third approach is statistical physics [1, pp. 14–16, pp. 184–195]. The energies used in these three treatments are essentially different from one another in characteristics [1, pp. 13–14, pp. 195–201] as discussed in Chap. 1. We have already noted in Sect. 2.1.4, which you will find later in Sect. 4.2.4, and pointed out that the difference of way treating the same problem substantially arises from placing concepts of different properties in the same category of energy just because they are all called energy.

2. **Uncritical mixing of quantum and classical mechanical energies**

In the absolute reaction kinetics, a number of autonomous / superficial combinations of quantum mechanical and classical mechanical concepts are introduced, to describe the phenomena of reaction rates. For example, in spite of the quantum mechanical approach to create the potential surface using London Eq. (2.2), the motion of atoms on the potential surface is treated completely under classical mechanics. The calculation of the mean velocity $\overline{v}$ of the activated complex of the formula (2.7) was also completely derived from the classical mechanical way. Basically, the motion of a mass point on the potential surface means a change in the arrangement of atoms, and the velocity of the motion of the mass point implies the velocity of configurational change of atoms. However, $\overline{v}$ in Eq. (2.7) directly represents the translational velocity of the activated complex as a whole in the translational motion. The surmount over a peak of the potential surface means that the configuration of atoms changes and the reactant becomes an activated complex and then a final product. Therefore, the key point is only the velocity of change in the arrangement of the atoms, but not the velocity of the translational motion of atomic cluster as a whole body without any change of atomic arrangement. If $\overline{v}$ in the Eq. (2.7) is the translational velocity of the activated complex without the rearrangement of atoms, this velocity is completely unrelated to the rate of chemical reaction. The treatment of diagonalization of kinetic energy [1, pp. 100–103] alone does not fully answer this question, and it is accordingly questionable whether such an explanation is possible for any multiparticle systems.

The second example of the mere autonomous/superficial coupling of quantum mechanics and classical mechanics is found in the mechanical non-logical separation of potential energy from kinetic energy in the motion of a particle on the potential surface. A chemical reaction at the temperature T is originally the comprehensive interaction among all electrons and nuclei of the related atoms. In *The Theory of Rate Processes* [1], the kinetic energy of the electrons and the Coulomb energy between nuclei and electrons are calculated quantum mechanically as eigenvalues in the eigenvalue equation, which are treated as potential energy. On the other hand, the kinetic energy of the nuclei is exclusively separated and treated as kinetic energy of the system [1, pp. 66–76, pp. 100–107]. It is again questionable whether such a treatment can be scientifically justified.

3. **Convertibility of the potential energy of electrons and the kinetic energy of atomic nuclei**

In *The Theory of Rate Processes* [1], the authors assumed that the sum of the potential energy and the kinetic energy is preserved, and the potential energy and the kinetic energy can be freely converted to each other in the motion of a mass point on the potential energy surface [1, pp. 100–107]. However, the potential energy is mainly related to the motion of electrons, and the kinetic energy is related to the motion of atomic nuclei. Recent studies of molecular collisions have revealed that the mutual energy conversion between the rotational, translational, and vibrational energy of molecules is different from each other and the conversion does not occur easily in real cases. Thus, when the hierarchies of the subjects of motion are also different, energy exchange or energy transfer between the different levels of hierarchies would be generally difficult.

4. **Question about the assumption that reactions proceed adiabatically**

Based on the absolute reaction kinetics, Eyring et al. calculated the activation energy on the potential surface, with the help of the adiabatic assumption. In other words, they assumed that there is no electron transition in the chemical reaction, and the reaction proceeds along one potential plane [1, pp. 2–3, p. 87]. However, recent studies have revealed that chemical reactions involve the excitation of internal motion of reactive molecules. Therefore, it is necessary to study the contribution of motion belonging to the hierarchy lower than the molecular level in more detail. Thus, physically strict consideration is needed to assume the activation energy as the potential energy difference between the reactant and the activated complex based on the adiabatic assumption.

2.1.6 Problems on Theoretical Consistency as Science

When considering the absolute reaction kinetics as a theoretical system of reaction rates, we can point out the following problems from the perspective described in Chap. 1.

1. **Viewpoint of a deductive theoretical system.**

For the establishment of the reaction kinetics, we must prepare a deductive theoretical system based on a few well-established fundamental laws. In this point, the absolute reaction kinetics lacks the fundamental rule for predicting the direction of change, as was already described in Sect. 2.1.3. The eigenvalue equation of quantum mechanics used in the theory is an equation that describes the stationary state as described in Sect. 1.8.1. Thermodynamics and statistical physics are both theoretical bases for describing equilibrium states. While reaction kinetics generally aims to describe a dynamic process that proceeds irreversibly in one direction, we have to refer that the description of absolute reaction kinetics is concerned, in principle, with reversible and static states in nature. Contrary to what should be a deductive theoretical system, the construction of the potential surface, for example, ultimately relies on the semi-empirical method with Morse formula, which is adjusted with the measurement data of molecular spectra [1, p. 2, pp. 91–100]. The formulation of the specific reaction rate itself is based on the empirical equation [1, p.13].

2. **Viewpoint of criteria of truthfulness in scientific cognition**

To understand the situation more clearly, we compared the absolute reaction kinetics with the criteria of truthfulness in scientific cognition? In Sect. 1.3, we pointed out the logical consistency and the experimental verifiability as criteria of truthfulness in scientific cognition. It has already been revealed from the discussion in the previous section that the absolute reaction kinetics never satisfies the logical consistency. Regarding the experimental verifiability, let us point out the following here. What Eyring's absolute reaction kinetics is characterized, in contrast to the Arrhenius' concept of active molecules, is the introduction of the concept of activated complexes. However, it is very difficult to directly observe whether such an activated complex eventually appears during the processes or not.

Thus, we summarize the fundamental problems included in the absolute reaction kinetics and discussed them from the viewpoints established in Chap. 1. In this chapter, the author has exclusively pointed out the shortcomings in the theory, and the present examination has revealed that this theory contained crucial and essential difficulties. However, the readers should not misunderstand the historically valuable achievements of the absolute reaction kinetics providing an overall perspective and useful physical insight with the rate processes of chemical reactions. Thus, though it is needless to say, we can use the discipline presented in the theory as a foothold.

2.2 Critique of Nucleation Theory

With a short glance, we find a closely related application of the absolute reaction kinetics, in the nucleation theory of phase transformation in metallic alloys. Since the nucleation theory was first proposed by Volmer and Weber [2] and Becker and Döring [3] for the condensation from vapor, the idea has been applied to all types of phase transformations. This section summarizes the basic features of the concept of nucleation and discusses its problems.

2.2.1 Basic Features in Concept of Nucleation in Precipitation

Under a temperature at which atoms are relatively mobile, local rearrangement of atoms occurs due to the diffusions associated with thermal fluctuations. In a supersaturated alloy, a high solute content embryos generated in the old phase have the possibility to grow into a nucleus of the new phase due to the diffusions of solute atoms. It is generally thought that when the embryo is formed, the change in free energy consists of three parts: (1) volume free energy, (2) interface energy, and (3) elastic strain energy. Since embryos are formed by thermal fluctuations, they may possibly vary in their (a) sizes, (b) shapes, and (c) compositions (in the case of multicomponent system). In addition, since internal structure of the embryo may also be inhomogeneous, uniformity can also be another parameter in the free energy estimation. Thus, the theory of nucleation is based on estimating the free energy change associated with the formation of the embryo as a function of these parameters.

The simplest example is as the following case where a liquid phase is formed from a gas phase. In this case, the strain energy among the abovementioned three energies can be ignored. This is because the strain energy becomes a problem only in the solid phase transformation, and in the present case, the solid phase is not involved. Let us further assume that the structure, composition, and properties of the embryo are the same as those of the large product phase grown at the end of the reaction, and that the inner structure of the embryo is homogeneous [4, Chap. 5]. Then, the energies related to the formation of the embryo are the volume free energy and the interfacial energy, and the shape of the embryo becomes spherical due to the requirement of the minimum interfacial energy. Therefore, the free energy change caused by the formation of the embryo is given by the following equation, where the spherical embryo of β phase (liquid phase) with a radius r is assumed to be formed in α phase (gas phase).

$$\Delta G = \frac{4}{3}\pi r^3 \cdot \Delta G_v + 4\pi r^2 \cdot \gamma \tag{2.12}$$

Here, ΔG_v represents the difference between the free energies measured in the large α phase and β phase per unit volume of the β phase. γ is the interfacial energy per unit area of the α phase–β phase interface and is assumed here to be independent on the crystallographic orientation relationship and the radius r. The first term of Eq. (2.12) is the volume free energy, and the second term is the interface energy. The second term is always positive. Under a certain temperature range where α is stable compared to β, the first term is also positive, so that ΔG becomes positive and increases sharply with r. ΔG is negative if β is stable compared to α. When r is a small value, the term relating to the interface energy is dominant, and ΔG is positive. However, for large r, the volume free energy term is dominant. ΔG has a maximum value of W at the radius r_c. The results are shown in Fig. 2.4. r_c and W are related with the equation:

$$\left(\frac{\partial \Delta G}{\partial r}\right)_{r=r_c} = 0 \tag{2.13}$$

Using Eq. (2.12)

$$r_c = -\frac{2\gamma}{\Delta G_v} \tag{2.14}$$

Therefore, we have

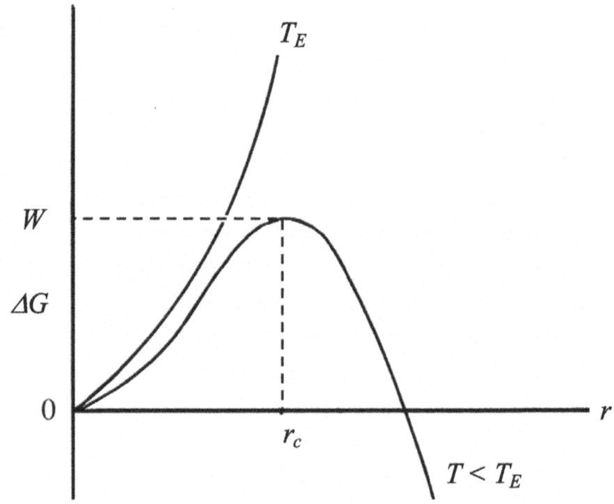

Fig. 2.4 Free energy of formation of spherical embryos as a function of the radius

$$W = \frac{16}{3}\pi\frac{\gamma^3}{(\Delta G_v)^2} \tag{2.15}$$

where r_c is the critical value and W is the activation free energy for nucleation.

At the equilibrium temperature T_E, $\Delta G_v = 0$, so the critical size and activation free energy become infinite. This situation means that at equilibrium temperature, no phase transformation takes place, and this speculation is consistent with such experimental facts as that undercooling or overheating is always required for the phase transformation to take place.

The distribution function of the embryos, i.e., the number of embryos in a unit volume as a function of the magnitude of the embryos, is then calculated. This calculation is performed based on the following assumptions.

(Assumption 1) The system is actually in equilibrium, and even if the nucleus with critical sizes is removed from the equilibrium distribution, the equilibrium distribution is not disturbed.

(Assumption 2) A group of the embryo of any sizes behaves independently with the group of the embryo of all other sizes. This assumption allows us to calculate the number of embryos of a given size independently of the number of embryos of the other size.

Suppose that there are N positions per unit volume where embryos can be formed, and there are n_r embryos of radius r per unit volume. The equilibrium between the N positions and the n_r embryos is described using the equilibrium constant K as follows:

$$\frac{n_r}{N} = K = \exp\left[-\left(\frac{\Delta G^0}{kT}\right)\right] \tag{2.16}$$

Substituting W of the Formula (2.15) into the standard free energy ΔG^0, the number of embryos having the critical size per unit volume is given by

$$n_c = N\exp\left[-\left(\frac{W}{kT}\right)\right] \tag{2.17}$$

Embryos with radius n_r become nuclei when one or more atoms are obtained. If the activation energy in the jumping process of passing through the interface is U_I the moving velocity of the interface is proportional to $\exp(-U_I/kT)$. Also, the frequency at which the embryos of critical sizes become stable is given by

$$n_s \cdot P \cdot \nu\exp\left[-\left(\frac{U_I}{kT}\right)\right] \tag{2.18}$$

where n_s is the number of atoms in the parent phase that is in contact with the surface of the embryos having critical size, and ν is the frequency of thermal vibration of these atoms. P is also the probability that the vibration of these atoms will occur in the direction of the embryo multiplied by a coefficient indicating that those atoms will adhere only to specific points on the surface of the embryo. From Eqs. (2.17) and (2.18), the number of nuclei I generated in a unit volume of the parent phase per unit time in the stationary state is given by

$$I = N \cdot n_s \cdot P \cdot \nu \exp\left[-\left(\frac{U_I}{kT}\right)\right] \cdot \exp\left[-\left(\frac{W}{kT}\right)\right] \tag{2.19}$$

$$= A \cdot \exp\left[-\left(\frac{U_I + W}{kT}\right)\right] \tag{2.20}$$

Frequency factor A is the product of the pre-exponential terms in Eq. (2.19).

The abovementioned flow of logical expansion is the basic idea of Volmer–Weber theory. In the Volmer–Weber theory, the premises that the composition, structure, and properties of embryos are kept constant are used, and accordingly the change in free energy associated with the formation of embryos is dependent on size only. Becker extended this idea to the transformations in binary systems. Borelius considers that the composition of embryo is changeable and the change in free energy accompanying the formation of the embryo is a function of composition alone. Hobstetter combined both of these models to simultaneously account for the size and composition of the embryos. Furthermore, Cahn and Hilliard treated the size, composition, and also heterogeneity as changeable variables, taking into account the heterogeneity inside the embryos. Nabarro claimed that the strain energy neglected in these treatments should be taken into account. Thus, the nucleation theory has been developed step by step, integrating various parameters, and the application has been expanded into the non-uniform nucleation at an incomplete structural area (lattice defect). However, the basic idea has been unchanged.

2.2.2 *Problems of Nucleation Theory*

In the previous section, the basic features of the concept of nucleation were briefly summarized. We will examine the issues of nucleation theory with the critique of the absolute reaction kinetics in this section.

1. **Scope of nucleation theory**

We can point out a problem regarding the scope of nucleation theory. Most of nucleation theory assumes the case where nucleation is caused by thermal fluctuation or thermal activation. However, the nucleation in phase transformation is not always due to thermal fluctuations or thermal activation. For example, martensitic transformation occurs at such lower temperatures and a high rate that cannot be attributed to thermal activation. Therefore, the phenomenon of phase transformation requires a more general and unified theory that can explain the nucleation not only due to thermal activation but also due to non-thermal activation.

2. **Assumption of a definite energy**

As for the assumption of a definite energy, in nucleation theory, (free) energy is considered as a single-valued function of a state, for example shown in Fig. 2.4. However, any state is not always a single-valued state of energy. Unstable transition states are characterized by indefinite energies and represented as superposition of different energies, as revealed later in Sect. 4.1.3 of Chap. 4.

3. **Fundamental law that gives the direction of change**

Regarding the fundamental law that gives the direction of change, the virtual situation in the case of nucleation theory seems to be quite different from the assumption used in the absolute reaction kinetics. In the absolute reaction kinetics, the sum of the potential energy and the kinetic energy was kept constant and reserved in the motion of a particle on the potential energy surface. Thus, the energy is the same in all states of the reactants, activated complexes, and products. Even if the ratio between the potential energy and the kinetic energy changes, the total energy is kept same; therefore, no preference for any state should appear. Consequently, the directionality of the reaction from the reactant to the product would not appear.

On the other hand, free energy is used in the nucleation theory. According to the fundamental law of thermodynamics, the spontaneous changes in an isolated system correspond to the decrease in free energy, and the phenomenon of increasing free energy does not occur. The nucleation theory, therefore, seems to involve a basic law that predicts the direction of change. However, we point out that this law of decrease in free energy prevents the nucleation from occurring. This is because the nucleation is accompanied by the increase in free energy. In addition, the free energy change does not include the concept of time. Therefore, this law alone does not help to describe the reaction rate.

4. **Assumption of equilibrium in the distribution of embryos**

In the nucleation theory, as mentioned in the previous section, the assumption of equilibrium was used for the distribution of embryos. That is, it is assumed that the process of embryo formation is in equilibrium, and that this equilibrium is not disturbed by the continuous removal of embryos with critical sizes. This assumption is exactly the same as the equilibrium assumption made between the reactant and the activated complex in the absolute reaction kinetics, as described in Sect. 2.1.4. Therefore, the conclusions reached there apply exactly to the nucleation theory. These conclusions can be summarized as follows in line with the theory of nucleation.

(1) Considering the whole reaction as composed of two processes, nucleation and growth, we find the situation under the nucleation process occurring much faster than the growth process. This situation, however, leads to such an irrational consequence that the nucleation process which must not be a rate-determining step controls the overall reaction rate.

(2) In the initial part of Volmer–Weber's theory, for example, free energy was expressed as a function of the radius of the embryo (Eq. (2.12)). This equation shows an increase in free energy accompanied by the formation of embryos.

Consequently, nucleation appeared to be a thermodynamically unexplainable phenomenon. In the later part of Volmer–Weber's theory, the process of embryo formation was assumed to be in equilibrium. Contrarily to the logical development in the first part, the process of embryo formation did not involve an increase in free energy. In other words, as shown in Fig. 2.2, the free energy does not change during the process of forming an embryo, and the activation energy corresponds to the standard free energy. However, the standard free energy is different from the activation energy and cannot be regarded as equivalent to the activation energy; this has already been described in Sect. 2.1.4. Thus, in Volmer–Weber's theory, W in Eq. (2.15) is substituted for ΔG^0 in Eq. (2.16), but this is logically unfounded. This is because ΔG^0 is the standard free energy and W is the free energy, which are completely different in character, and the physical meanings given to the terms are also different.

(3) The theoretical treatment of the nucleation process seems to become extremely strange. With no assumption of equilibrium in the process of embryo formation, the free energy would increase during the progress of this process, and this situation means that the formation of embryos cannot occur because of the fundamental laws of thermodynamics. Therefore, despite an attempt to describe the process of nucleation in terms of free energy for the thermodynamic explanation, the description resulted in failure, contrary to expectations. Assuming equilibrium in the process of embryo formation, on the other hand, there is no free energy peak that should be regarded as the cause of the difficulty in nucleation. The peak of standard free energy cannot be considered as activation energy. In addition, such a treatment in the process of embryo formation forces the left and right sides of the critical size of embryo to be completely different characters. Without the equilibrium assumption, the left of the critical size of embryos is a stage that does not obey the laws of thermodynamics, and the right is the stage of thermodynamics. The formation and growth of nuclei should be homogeneous and continuous change in nature. However, the treatment described above introduces a heterogeneous and discontinuous change at the critical size of the embryos, resulting in the separation of the two inconsistent stages. One of the causes of such confusion is due to the introduction of an unreasonable equilibrium assumption, as described in Sect. 2.1.4.

5. **Problems of nucleation theory from a hierarchical viewpoint**

Let us consider the problems of nucleation theory from a hierarchical viewpoint. A distinctive feature emerged at the comparison of the nucleation theory with the absolute reaction kinetics is that the nucleation theory is purely based on a macroscopic viewpoint. In the absolute reaction kinetics, the treatment from a microscopic viewpoint was dominant. In other words, the London equation (2.2) used to create the potential energy surface was derived quantum mechanically. Since the treatment of absolute reaction kinetics is based on this potential energy surface, the viewpoint is basically microscopic. In contrast, the nucleation theory is described the phenomena using thermodynamics and therefore the hierarchical viewpoint is basically macroscopic. We will consider below whether we can describe the nucleation process using a macroscopic perspective?

(1) Eq. (2.12) consists of volume free energy and surface energy terms. More generally, strain energy is calculated in addition to these two energies, in the nucleation theory. The terms are all derived from macroscopic concepts. For example, we cannot define volume or surface of one atom or one electron. In this case, therefore, the volume free energy or surface energy cannot be defined. They can be defined only for macroscopic systems. In terms of the strain energy, we also fail to define for a system consisting of one atom, and then the strain energy can be considered only for a macroscopic object consisting of a huge number of atoms. Thus, all the concepts used in the description of nucleation are macroscopic concepts. For the process of nucleation itself, on the other hand, we must describe the formation of embryos from one atom and one electron. Therefore, by no means the macroscopic concepts used in the nucleation theory are properly given in the microscopic nature of the phenomena they deal with. The logically consistent treatment of nucleation is not thus provided with macroscopic, but microscopic discipline. Furthermore, the consideration with the concepts of nucleation itself is only valid in a macroscopic treatment, and accordingly the basis for the concepts of nucleation loose the consistency in the original microscopic treatment.
(2) The predominant assumption is intuitively used in Eq. (2.12) that the calculation of the free energy associated with the formation of the embryo is given by the sum of the volume free energy and the surface energy. Such arithmetic additivity of thermodynamic quantities, as well as the additivity of physical quantities in general is justified only when the system is macroscopic and therefore almost isolated, as described in Sect. 1.8.3. Thus, not only the concepts of volume free energy or surface energy does not hold but also the basis of their additivity is lost, in the case that the system is not macroscopic as in nucleation.
(3) Next issue that we need consider is the problem whether nucleation theory can be described by thermodynamics or not. As already mentioned in Sect. 1.8.3, thermodynamics is a logical system that originally deals with equilibrium states of temperature-related phenomena in macroscopic systems consisting of a huge number of particles and formulates the necessary conditions for such phenomena. From this point of view, the systems involved in nucleation are microscopic systems consisting of a small number of particles. In addition, the process of nucleation is a progressive change that occurs only because the system is not in equilibrium. Thus, thermodynamics is not an appropriate logical means to describe nucleation phenomena. Considering the second thermodynamic law of increasing entropy which represents irreversibility and the law of decreasing free energy which is another expression of the law of increasing entropy, we understand that thermodynamic description can be applied only to macroscopic systems consisting of huge number of particles, neither the law can be used to explain the irreversible change in the nucleation process. We have already seen

that the law applied to the process of nucleation leads to the conclusion that nucleation is impossible to occur. Thus, thermodynamics is, by its very nature, unsuitable for describing nucleation.

6. **Problems in the nucleation theory from theoretical viewpoint**

Let us discuss some problems that arise when the nucleation theory is considered from theoretical point of view. Just like the absolute reaction kinetics, the nucleation theory has the following problems.

(1) The purpose of nucleation theory is to describe microscopic dynamic processes that proceed irreversibly in one direction. On the other hand, the thermodynamics is a macroscopic logical system that describes an equilibrium state. Accordingly, thermodynamics is not suitable for the purpose of describing the process of nucleation. Therefore, the law of decreasing free energy and the law of increasing entropy, which seems to serve as the fundamental laws giving the direction of change at first glance, are ultimately revealed to be useless. Moreover, we have also learned that the inappropriate use of thermodynamics to describe the nucleation process brought fundamental difficulties in the consideration. Thus, judging by the criteria for a theoretical system, *i.e.*, a deductive system of logic based on a few well-established fundamental laws, we must conclude that the conventional nucleation theory is ceased in a highly unsatisfactory attainment.

(2) According to the criterion of scientific justification, following problems also arise in the nucleation theory. With respect of logical consistency, the serious defect has already been pointed out that logical inconsistency exists in nucleation theory. Regarding the possibility of experimental verification, nobody has ever succeeded to directly identify the "critical nucleus", the most important concept in nucleation theory. Not only that, but also the method of directly observing critical nuclei has not yet been established. Thus far, the nucleation theory, which is closely related to the absolute reaction kinetics and can be considered as its application, has been discussed in order to clarify the fundamental problems, corresponding to the items taken up in the case of the absolute reaction kinetics. As a result, it was found that the inherent difficulties of the absolute reaction kinetics are directly taken over in the nucleation theory. Thus, the fundamental problems involved in the absolute reaction kinetics and the nucleation theory become today's issue in the reaction kinetics that we need get further understandings. In the next chapter, we will examine the details of the analysis in this chapter, discuss the alternative scheme of the reaction kinetics, and develop a new theory of reaction rates to overcome fundamental problems in the conventional interpretations.

References

1. S. Glasstone, K.J. Laidler, H. Eyring, *The Theory of Rate Processes* (McGraw-Hill Book Company, 1941)
2. M. Volmer, A. Weber, Nucleus formation in supersaturated systems. Z. Phys. Chem. **119**, 277–301 (1926)
3. R. Becker, W. Döring, Kinetische behandlung der keimbildung in ubersattigten dampfern Annalen Der Physik **24**, 719–752 (1935). https://doi.org/10.1002/andp.19354160806
4. J. Burke, *The Kinetics of Phase Transformations in Metals* (Pergamon Press, Oxford, 1965)

Further Readings

The Chemical Society of Japan, Chemistry Sourcebooks, No.6, *Chemical Reactions*, [in Japanese] (Academic Publishing Center, Tokyo, 1976)

H. Eyring, E.M. Eyring, *Modern Chemical Kinetics* (Reinhold Publishing, 1963)

K. Hirota, H. Arai, S. Tsuchiya, Y. Ookatsu, T. Asaba, *Reaction Rate*, [in Japanese] (Kyoritsu Publishing, Tokyo, 1964)

I. Amdur, G.G. Hammes, *Chemical Kinetics* (McGraw-Hill Book Company, 1966)

T. Keii, *Chemical Reaction Rate Theory*, 3rd edn. [in Japanese] (Tokyo-Kagaku Doujin, Tokyo, 2001)

R.D. Levin and R.B. Bernstein, *Molecular Reaction Dynamics* (Oxford University Press,1974)

J.W. Christian, *The Theory of Transformation in Metals and Alloys* (Pergamon Press, 1965)

Part II
Formulation of Our Reaction Kinetics

Chapter 3
Physical Formulation of Our Theory

Abstract This chapter describes the physical formulation of reaction kinetics newly developed by the present author (S. Yamamoto), using Schrödinger's wave equation, uncertainty principle, and basic concepts already established in quantum mechanics. The author has specially stressed that the eigenvalues of energy in stationary state are not suitable to consider the kinetics of irreversible reaction, but the fluctuations of the energy are essential, instead. The detailed explanation of the new formulation has been performed, throughout based on the quantum mechanics, but without semi-classical notions which were inconsistently glided into the conventional reaction kinetics. The fluctuation of energy, one of the crucial concepts, is also justified. The validity of our reaction kinetics has been examined in wide-range examples of physical and chemical reactions, and the applicability of the present theory has been shown in this chapter.

Keywords Theoretical foundation · Quantum mechanics · Wave equation · Transition state · Fluctuation of energy · Lifetime

We have discussed the intrinsic meaning of reaction kinetics in a scientific cognition in Chap. 1, and further in Chap. 2, we clarified the crucial problems involved in the current reaction kinetics and phase transformation kinetics described in Chap. 1 from the fundamental viewpoints. In this chapter, we would present an attempt of description of reaction kinetics along the guideline for constructing a new theory shown in the previous chapters.

As already mentioned, in Chap. 2, the crucial problems in the current theoretical systems of reaction kinetics and phase transformation kinetics are mingled: We first raise the assumption of definiteness in energy for dealing with transition states, second, the absence of a fundamental law that predicts the direction of change, and third, the assumption of equilibrium in transition states.

S. Yamamoto, *Reaction Kinetics Based on Time-Energy Uncertainty Principle*,
https://doi.org/10.1007/978-981-19-9673-3_3

Although we can find many other problems, these three issues seem to be quite significant when considering the feasibility of reaction kinetics. Thus, we will start to discuss about three issues in the first part of this chapter and finally lead our consensus to the following conclusions:

(1) Quantum mechanical uncertainty is important in transition states.
(2) The uncertainty principle, which quantitatively expresses the uncertainty in transition states, is useful as a fundamental law that predicts the direction of change.
(3) This principle enables us to release the equilibrium assumption in transition states.

3.1 Two Approaches Related to Transition States

We shall first show that the transition state is described as a state of quantum mechanical uncertainty and that the quantitative expression of the quantum mechanical uncertainty is related to the time–energy uncertainty relation $\Delta t \cdot \Delta E \cong \hbar$. We can grasp the physical insight to the uncertainty when a particle is passing through a potential wall in Sect. 3.1.1 and the uncertainty when the perturbing potential is applied to a system in Sect. 3.1.2 [1, Chaps. 8, 11, 18].

3.1.1 Particles Traveling Through a Square Potential

At first, let us consider a simple example of a particle traveling in one-dimensional direction with a square-shaped potential. Using Eq. (1.1) noted in Chap. 1, we can describe the wave equation of a quantum system as

$$i\hbar\frac{\partial}{\partial t}\Psi = H\Psi \tag{3.1}$$

For simplicity, we here suppose that Hamiltonian H does not explicitly contain time. Let E express a constant energy, and ψ a time-independent wave function that is only related to the coordinates of the configuration space. We assume Ψ:

$$\Psi = \psi e^{-iEt/\hbar} \tag{3.2}$$

Substituting Eq. (3.2) into Eq. (3.1), we obtain the eigenvalue equation,

$$H\psi = E\psi \tag{3.3}$$

Equation (3.3) is called the time-independent wave equation. The wave function Ψ relates to a spatial distribution of electrons in a stationary state. We can obtain wave function Ψ by multiplying the phase factor $\exp(-iEt/\hbar)$ to time-independent wave function ψ. Now, we solve the wave Eq. (3.3) under time-independent potential field V, referring to

$$H = \frac{\hbar^2}{2m}\frac{\partial^2}{\partial x^2} + V \tag{3.4}$$

Equation (3.3) transforms into

$$-\frac{\hbar^2}{2m}\frac{\partial^2}{\partial x^2}\psi = (E - V)\psi \tag{3.5}$$

The solution of Eq. (3.5) is given as follows:

$$\psi = A\exp\left\{i\sqrt{2m(E-V)}\left(\frac{x}{\hbar}\right)\right\} + B\exp\left\{-i\sqrt{2m(E-V)}\left(\frac{x}{\hbar}\right)\right\} \tag{3.6}$$

where A and B are arbitrary constants. Time-dependent solution is given by

$$\Psi = A\exp\left\{\frac{i(Px - Et)}{\hbar}\right\} + B\exp\left\{-\frac{i(Px + Et)}{\hbar}\right\} \tag{3.7}$$

where

$$P = \sqrt{2m(E - V)}$$

The first term of Eq. (3.7) represents a wave moving to the right, and the second term represents a wave to the left.

Let us now consider a square-shaped potential model shown in Fig. 3.1. The potential in Fig. 3.1a has a constant value of V_0 in the region between $x = -a$ and $x = a$, and 0 for the other region (If V_0 is negative as in Fig. 3.1b, the following equations can also be applied.).

Suppose that the particle travels from left to right. According to the wave theory, the traveling particle (*i.e.*, wave) is reflected by the potential wall at $x = -a$ and $x = a$. As a result, the reflected and transmitted waves give rise. To treat this problem, we shall consider the phenomenon occurring at the region $x > a$ where only the transmitted wave is present. In this region, we obtain the wave function as

$$\psi = A\exp\left(\frac{iP_1x}{\hbar}\right) \tag{3.8}$$

Here, P_1 is the momentum of the particle in the region, and $P_1 = \sqrt{2mE}$ under non-relativistic scheme.

In the region of the potential peak between $x = -a$ and $x = a$, the wave function is expressed by

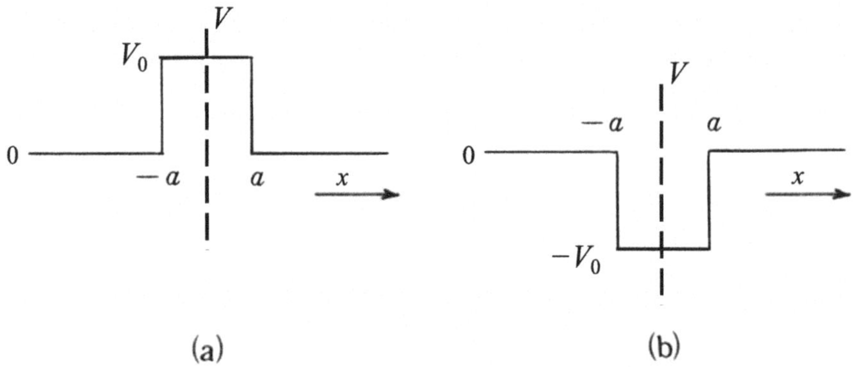

Fig. 3.1 One-dimensional square-shaped potential with a finite height/depth

$$\psi = B\exp\left(\frac{iP_2x}{\hbar}\right) + C\exp\left(-\frac{iP_2x}{\hbar}\right) \tag{3.9}$$

where

$$P_2 = \sqrt{2m(E - V_0)}$$

On the other hand, the wave function in the $x < -a$ region is given by

$$\psi = D\exp\left(\frac{iP_1x}{\hbar}\right) + F\exp\left(-\frac{iP_1x}{\hbar}\right) \tag{3.10}$$

Using the condition that ψ and $\mathrm{d}\psi/\mathrm{d}x$ are both continuous between $x = -a$ and $x = a$, the coefficients B, C, D, and F are given by following equations using A. That is,

$$B = \frac{A}{2}\left(1 + \frac{P_1}{P_2}\right)\exp\left\{\frac{i(P_1 - P_2)a}{\hbar}\right\} \tag{3.11}$$

$$C = \frac{A}{2}\left(1 - \frac{P_1}{P_2}\right)\exp\left\{\frac{i(P_1 + P_2)a}{\hbar}\right\} \tag{3.12}$$

$$D = \left(\frac{A}{4}\right)\exp\left(\frac{2iP_1a}{\hbar}\right)\left[\left(1 + \frac{P_2}{P_1}\right)\left(1 + \frac{P_1}{P_2}\right)\exp\left(-\frac{2iP_2a}{\hbar}\right) + \left(1 - \frac{P_2}{P_1}\right)\left(1 - \frac{P_1}{P_2}\right)\exp\left(\frac{2iP_2a}{\hbar}\right)\right] \tag{3.13}$$

$$F = \left(\frac{A}{4}\right)\left[\left(1 - \frac{P_2}{P_1}\right)\left(1 + \frac{P_1}{P_2}\right)\exp\left(-\frac{2iP_2a}{\hbar}\right)\right.$$

$$+\left(1+\frac{P_2}{P_1}\right)\left(1-\frac{P_1}{P_2}\right)\exp\left(\frac{2iP_2a}{\hbar}\right)\Bigg] \tag{3.14}$$

Calculation using these equations, the transmittance (i.e., the ratio of the intensities of the incident and transmitted waves) is obtained as following,

$$T=\frac{|A|^2}{|D|^2}=\frac{1}{1+\frac{1}{4}\left(\frac{P_1}{P_2}-\frac{P_2}{P_1}\right)^2\sin^2\left(\frac{2P_2a}{\hbar}\right)} \tag{3.15}$$

This equation is worth examining in detail. Using Eq. (3.15), we know $T = 1$ for $P_1 = P_2$. In this case, this result is completely natural since there is no potential change. If $P_1 \neq P_2$, the transmittance is generally less than unity, which means that the reflection occurs. However, there is still a case where $T = 1$ even for $P_1 \neq P_2$.

This occurs when $\sin^2(2P_2a/\hbar) = 0$, i.e.,

$$P_2=\frac{N\pi\hbar}{2a}\quad (N:\text{integer}) \tag{3.16}$$

What happens when this condition is satisfied? The wave is reflected at $x = -a$ and $x = a$, so the wave in the potential barrier goes back and forth between $x = -a$ and $x = a$. The phase shift that the wave undergoes while crossing the potential barrier once and coming back is $2\pi \cdot 4a/\lambda = 4P_2a/\hbar$, which is equal to $2N\pi$ from Eq. (3.15). In other words, the waves that are reflected backward and run forward in the potential barrier and the waves that are just coming in are in the same phase, and there is a resonance that interferes with each other, resulting in strengthening each other. Thus, at a certain wavelength (momentum), the transmittance becomes one. The transmittance is dependent on the wave number $k(= P/\hbar = 2\pi/\lambda)$ as shown in Fig. 3.2. To estimate the widths of the peaks in Fig. 3.2, we assume that P_1/P_2 is large. Then, sharp peaks will appear periodically.

Next, we investigate the distance of the position where T is less than 1/2 from $P_2 = N\pi\hbar/2a$. We expect this situation happens in the case that the following condition is satisfied:

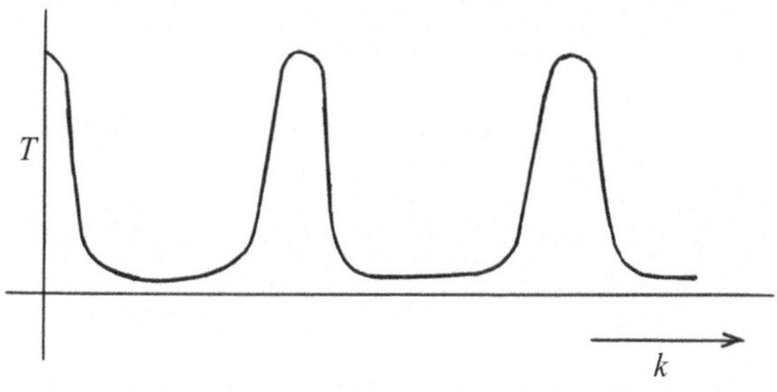

Fig. 3.2 Variation of transmittance T with wave number k [1]

$$\frac{1}{4}\left(\frac{P_1}{P_2}-\frac{P_2}{P_1}\right)^2\sin^2\left(\frac{2P_2a}{\hbar}\right)=1 \tag{3.17}$$

or

$$\sin\left(\frac{2P_2a}{\hbar}\right)=\pm\frac{2}{|P_1/P_2-P_2/P_1|} \tag{3.18}$$

If the denominator on the right-hand side becomes larger, then $2P_2a/\hbar$ is only slightly different from $N\pi$ in the case, and so the condition can be written as follows:

$$\frac{2P_2a}{\hbar}\cong N\pi\pm\frac{2}{|P_1/P_2-P_2/P_1|} \tag{3.19}$$

$$P_2\cong\frac{N\pi\hbar}{2a}\pm\frac{\hbar/a}{|P_1/P_2-P_2/P_1|} \tag{3.20}$$

$$\delta P_2\cong\frac{\hbar/a}{|P_1/P_2-P_2/P_1|} \tag{3.21}$$

Using the expression:

$$P_2=\sqrt{2m(E-V_0)}$$

and differentiating this formula, we can draw the following equations (we assume that $\delta P_2/P_2$ is small):

$$\delta P_2\cong\sqrt{\frac{m}{2(E-V_0)}}\delta E \tag{3.22}$$

and

$$\begin{aligned}\delta E&\cong\sqrt{\frac{2(E-V_0)}{m}}\delta P_2\cong\sqrt{\frac{2(E-V_0)}{m}}\left(\frac{\hbar}{a}\right)\frac{1}{|P_1/P_2-P_2/P_1|}\\&=\frac{v_2\hbar/a}{|P_1/P_2-P_2/P_1|}\end{aligned} \tag{3.23}$$

Here, v_2 is the velocity of the particles in the potential barrier.

For $P_1\gg P_2$

$$\delta E\cong v_2\frac{P_2}{P_1}\cdot\frac{\hbar}{a} \tag{3.24}$$

Let us now describe the traveling of a wave packet as a function of time in order to see how the influence of the potential appears on the motion. Starting with the region $x = -a$, we obtain the following equation from Eqs. (3.10) and (3.7) for time-dependent solutions:

$$\Psi(P_1) = \left[D(P_1)\exp\left(\frac{iP_1x}{\hbar}\right) + F(P_1)\exp\left(-\frac{iP_1x}{\hbar}\right)\right]\exp\left[-\frac{iE(P_1)t}{\hbar}\right] \quad (3.25)$$

where D and F are generally functions of P_1, and $E(P_1) = P_1^2/2m$. To create a wave packet, we need to multiply the weight factor $f(P_1 - P_0)$ to the peaks near a certain value, labeled P_0, and integrate $\Psi(P_1)$ with P_1, namely

$$\Psi(x,t) = \int \mathrm{d}P_1 f(P_1 - P_0)\exp\left[-\frac{iE(P_1)t}{\hbar}\right]$$
$$\left[D(P_1)\exp\left(\frac{iP_1x}{\hbar}\right) + F(P_1)\exp\left(-\frac{iP_1x}{\hbar}\right)\right] \quad (3.26)$$

Generally, D and F are smooth functions of P_1 and are defined by Eqs. (3.13) and (3.14). For convenience, we can choose the parameter A so that it satisfies $D(P_1) = 1$. This accordingly gives the following results:

$$\frac{A}{4} = \exp\left(-\frac{2iP_1a}{\hbar}\right)\left[\left(1+\frac{P_2}{P_1}\right)\left(1+\frac{P_1}{P_2}\right)\exp\left(-\frac{2iP_2a}{\hbar}\right)\right.$$
$$\left. + \left(1-\frac{P_2}{P_1}\right)\left(1-\frac{P_1}{P_2}\right)\exp\left(\frac{2iP_2a}{\hbar}\right)\right]^{-1} \quad (3.27)$$

With slight change of the form, we obtain

$$F = -i(P_1^2 - P_2^2)\exp\left(-\frac{2iP_1a}{\hbar}\right)\sin\left(\frac{2P_2a}{\hbar}\right)\left[2P_1P_2\cos\left(\frac{2P_2a}{\hbar}\right)\right.$$
$$\left. + i(P_1^2 + P_2^2)\sin\left(\frac{2P_2a}{\hbar}\right)\right]\cdot\left[4P_1^2P_2^2 + (P_1^2 - P_2^2)^2\sin^2\left(\frac{2P_2a}{\hbar}\right)\right]^{-1}$$
$$(3.28)$$

This equation is conveniently written as

$$F(P_1) = R(P_1)e^{-i\varphi_1} \quad (3.29)$$

where $R(P_1) = |F(P_1)|$. Taking the relation that P_2 is represented by P_1 into account, we reach

$$\varphi_1 = \frac{2P_1a}{\hbar} + \tan^{-1}\left[\frac{2P_1P_2}{P_1^2 + P_2^2}\cot\left(\frac{2P_2a}{\hbar}\right)\right] \quad (3.30)$$

Substituting this result into Eq. (3.26) for a further revised expression of Ψ,

$$\Psi(x,t) = \int \mathrm{d}P_1 f(P_1 - P_0)\exp\left[-\frac{iE(P_1)t}{\hbar}\right] \cdot \left\{\exp\left(\frac{iP_1x}{\hbar}\right) + R(P_1)\exp\left[-i\left(\frac{P_1x}{\hbar} + \varphi_1(P_1)\right)\right]\right\} \tag{3.31}$$

is obtained.

In order to consider the situations where Ψ takes the largest value and to find out where the phase of the wave has the extreme values, we differentiate this equation with P_1. This procedure ensures that many waves with different P_1 can be added in the same phase, resulting in peaks. For an incident wave, the phase has an extreme value when the following equation is satisfied,

$$\frac{\partial}{\partial P_1}\left(P_1\frac{x}{\hbar} - E(P_1)\frac{t}{\hbar}\right)_{P_1=P_0} = 0 \tag{3.32}$$

or

$$x = \left(\frac{\partial E}{\partial P_1}\right)_{P_1=P_0} \cdot t = \frac{P_0}{m}t \tag{3.33}$$

This maximum point moves infinitely to the left when $t \to -\infty$. Since the incident wave function has a physical meaning only when x is negative, it is obvious that the incident wave completely disappears after $t = 0$.

Let us now focus on the reflected wave. For realizing the extreme conditions in phase, the following equation should be satisfied.

$$\frac{\partial}{\partial P_1}\left(P_1\frac{x}{\hbar} + \varphi_1(P_1) + E(P_1)\frac{t}{\hbar}\right)_{P_1=P_0} = 0 \tag{3.34}$$

$$\begin{aligned} x &= -\left(\frac{\partial E}{\partial P_1}\right)_{P_1=P_0} \cdot t - \hbar\left(\frac{\partial \varphi_1}{\partial P_1}\right)_{P_1=P_0} \\ &= -\frac{P_0}{m}t - \hbar\left(\frac{\partial \varphi_1}{\partial P_1}\right)_{P_1=P_0} = -v_0 t - \hbar\left(\frac{\partial \varphi_1}{\partial P_1}\right)_{P_1=P_0} \end{aligned} \tag{3.35}$$

According to Eq. (3.35), we see that $x \to -\infty$ when $t \to +\infty$. Therefore, the reflection of wave packet appears after the incident wave packet arrives at the potential barrier. The significance of terms containing $(\partial\varphi_1/\partial P_1)$ in Eq. (3.35) will be discussed later.

The amplitude of the transmitted wave is expressed as

$$\Psi(x,t) = \int \mathrm{d}P_1 f(P_1 - P_0)\exp\left[-\frac{iE(P_1)t}{\hbar}\right] A(P_1)\exp\left(\frac{iP_1x}{\hbar}\right) \tag{3.36}$$

Representing this equation similarly to Eq. (3.29) and taking Eq. (3.27) into account, we obtain

$$A = |A|e^{i\varphi_2} \tag{3.37}$$

$$A = \frac{\exp(-2iP_1 a/\hbar)[\cos(2P_2 a/\hbar) + (i/2)(P_1/P_2 + P_2/P_1)\sin(2P_2 a/\hbar)]}{\cos^2(2P_2 a/\hbar) + (1/4)(P_1/P_2 + P_2/P_1)^2 \sin^2(2P_2 a/\hbar)} \tag{3.38}$$

The phase φ_2 is given by

$$\varphi_2 = -\frac{2P_1 a}{\hbar} + \tan^{-1}\left[\frac{1}{2}\left(\frac{P_1}{P_2} + \frac{P_2}{P_1}\right)\tan\left(\frac{2P_2 a}{\hbar}\right)\right] \tag{3.39}$$

With Eqs. (3.38) and (3.39), the wave function is described as

$$\Psi(x,t) = \int \mathrm{d}P_1 f(P_1 - P_0)|A|\exp\left\{i\left[P_1\frac{x}{\hbar} + \varphi_2 - E(P_1)\frac{t}{\hbar}\right]\right\} \tag{3.40}$$

The maximum of the wave packet appears at the x value where the derivative in the exponential function becomes zero. Namely, the condition is identified to be

$$x = \left(\frac{\partial E}{\partial P_1}\right)_{P_1=P_0} \cdot t - \hbar\left(\frac{\partial \varphi_2}{\partial P_1}\right)_{P_1=P_0} = \frac{P_0}{m}t - \hbar\left(\frac{\partial \varphi_2}{\partial P_1}\right)_{P_1=P_0} \tag{3.41}$$

When $t \to +\infty$, the maximum value appears in the region $x > a$. After sufficient progress of time, the transmitted wave appears and proceeds with the group velocity of $v_0 = P_0/m$.

If there is no potential barrier that would create a reflected wave, then the moving of transmitted wave with its center at $x = P_0 t/m$ should be expected. The additional term in Eq. (3.41) represents the time delay. Therefore, the influence of potential term appears as the result that x value with the second term is smaller than the one without the term in Eq. (3.41). The time delay Δt due to the potential barrier is directly drawn from Eq. (3.41). The result is shown by the following relation,

$$\Delta t = \frac{\hbar}{v_0}\left(\frac{\partial \varphi_2}{\partial P_1}\right)_{P_1=P_0} \tag{3.42}$$

We need to take the following condition into account to differentiate φ_2.

$$P_2^2 = 2m(E - V_0) = P_1^2 - 2mV_0 \tag{3.43}$$

Therefore,

$$P_2\left(\frac{\partial P_2}{\partial P_1}\right) = P_1, \frac{\partial P_2}{\partial P_1} = \frac{P_1}{P_2} \tag{3.44}$$

Then, we obtain

$$\hbar\left(\frac{\partial \varphi_2}{\partial P_1}\right) = -2a\left(\frac{P_1}{P_2}\right) + \frac{\hbar}{2}\left\{\left(\frac{2}{P_2} - \frac{P_2}{P_1^2} - \frac{P_1^2}{P_2^3}\right)\tan\left(\frac{2P_2a}{\hbar}\right)\cdot\right.$$
$$\left. + \left(\frac{2a}{\hbar}\right)\left[\left(\frac{P_1}{P_2}\right)^2 + 1\right]\sec^2\left(\frac{2P_2a}{\hbar}\right)\right\}\cdot\left[1 + \frac{1}{4}\left(\frac{P_1}{P_2} + \frac{P_2}{P_1}\right)^2 \tan^2\left(\frac{2P_2a}{\hbar}\right)\right]^{-1} \tag{3.45}$$

In the case of $P_1 = P_2$ (no potential walls), we can see that $\Delta t = 0$. When $P_1 \neq P_2$, waves (i.e., quantum particles) are reflected by potential walls, so that waves are reflected backward and running forward inside their potential walls. This should make Δt positive. This effect will be stronger in the vicinity of resonance transmission. This particularly holds for $P_1 \gg P_2$ because the reflectance is very large.

We shall consider the computation of Δt. For the resonance transmission, we refer to the next equations:

$$\left.\begin{array}{l} \tan\left(\frac{2P_2a}{\hbar}\right) = 0 \\ \text{and} \\ \sec^2\left(\frac{2P_2a}{\hbar}\right) = 1 \end{array}\right\} \tag{3.46}$$

Then, we obtain

$$v_0\Delta t = a\left[\left(\frac{P_1}{P_2}\right)^2 - 1\right] \tag{3.47}$$

In the case of $P_1 \gg P_2$, the time delay Δt is positive

$$\Delta t = \frac{a}{v_0}\left(\frac{P_1}{P_2}\right)^2 \tag{3.48}$$

where v_0 is the velocity of particles outside the potential well. From Eq. (3.48) for defining Δt and Eq. (3.24) for defining resonance width ΔE, we obtain (note that $v_1 = v_0$ in this case)

$$\Delta t \cdot \Delta E \cong \hbar \tag{3.49}$$

We can now carry out our calculations of uncertainties in the case that particles pass through the square-shaped potential, but it would be worthwhile to mention that similar discussions can be carried out in the following cases:

(1) The same treatment can be applied for a well-shaped potential model as shown in Fig. 3.1b, by setting V_0 sign to a negative value.
(2) The opposite case in which the potential energy changes very slowly as a function of position, the so-called Wentzel-Kramers-Brillouin (WKB) approximation can be applied and, in this case, similar computed results as described above will be obtained [2, Chap. 6].

Before closing the present consideration, we shall touch some physical implications of the results obtained above. The abovementioned results show that even when a particle has enough energy to escape, after entering the potential, the particle is reflected many times back and running forth before it manages to escape from the potential. If the number of the wave reflection is very large, the system will appear almost stationary. However, it gradually collapses as the wave passes slowly through the system after internal multiple reflections. Such a state is called a virtual state or metastable state. A particle in this state has positive energy and reveals contrast difference from the true bound state which has negative energy. The lifetime of the metastable state is given by Δt, which is calculated by Eq. (3.48). The wave function of such a metastable state creates a wave packet that passes through the potential in the time Δt. In order to create such a wave packet, energy in the range of ΔE is required. On the other hand, when the virtual bound state is broken, a particle having energy in this range appears. The uncertainty relation (3.49) is satisfied between the lifetime Δt of the virtual state and the energy fluctuation width ΔE in the virtual state [2, Chap. 8].

3.1.2 Perturbation and Uncertainty State

In this section, let us perform a calculation using the perturbation method, which is another approach to show that the transition state is under a condition of quantum mechanical uncertainty. Then, we will show that a system, which is in a definite state before the perturbation is turned on, becomes an uncertain state after the perturbation is applied. In the perturbation method, we start with a system for which the wave equation can be solved exactly and then investigate what happens when a small external disturbance acts on this system. To apply this method, we start with the following wave equation:

$$i\hbar\frac{\partial \Psi}{\partial t} = H\Psi \tag{3.50}$$

In the present treatment based on the perturbation theory, we confine our consideration in the case that Hamiltonian consists of two terms as

$$H = H_0 + \lambda V(x, p, t) \tag{3.51}$$

In the discussion, we assume that H_0 is a Hamiltonian operator of an unperturbed system whose eigenvalues and eigenfunctions are known. On the other hand, λV is a small perturbation term, and the coefficient λ is a constant that represents the strength of the perturbation. If λ is small enough, that is, if the perturbation is weak, the solution of the wave equation will not be very different from the solution obtained when $\lambda = 0$. In the case of the unperturbed state with $\lambda = 0$, the solution should be expanded to the series of eigenfunctions of H_0, $U_n(\boldsymbol{x})\exp\left[-iE_n^0t/\hbar\right]$, where E_n^0 is the n-th eigenvalue of H_0, and we obtain

$$\psi_{\lambda=0} = \sum_n C_n U_n \exp\left[-\frac{iE_n^0 t}{\hbar}\right] \tag{3.52}$$

The set of arbitrary constants C_n is determined by the boundary condition.

It should be noted that the method used to find the solution when $\lambda \neq 0$ generally implies that the solution $\psi(x)$ can be expressed as the series of $U_n(\boldsymbol{x})$ given at any time. The coefficients of $U_n(\boldsymbol{x})$ should also be a function of time, since the function changes with time. If the coefficients take a distinctive time variation $C_n \exp\left[-iE_n^0t/\hbar\right]$ and C_n is a constant, the series should be the solution of the wave equation $i\hbar(\partial\psi/\partial t) = H_0\psi$ for an unperturbed system. More generally, since these coefficients depend on time in a more complex manner, the wave function is expressed by,

$$\psi_{\lambda\neq 0} = \sum_n C_n \exp\left[-\frac{iE_n^0 t}{\hbar}\right] U_n(\boldsymbol{x}) \tag{3.53}$$

In this case, C_n is no longer constant and depends on time. For this reason, the above method is called the "method of variation of constants" based on mathematical terminology.

Substituting the above series (3.53) [with time-dependent C_n] into the wave Eq. (3.50) to find the solution, we then obtain

$$\begin{aligned}
&\sum_n \left(i\hbar\dot{C}_n + E_n^0 C_n\right) U_n(\boldsymbol{x}) \exp\left[-\frac{iE_n^0 t}{\hbar}\right] \\
&= \sum_n E_n^0 \exp\left[-\frac{iE_n^0 t}{\hbar}\right] U_n(\boldsymbol{x}) C_n + \lambda \sum_n V C_n U_n(\boldsymbol{x}) \exp\left[-\frac{iE_n^0 t}{\hbar}\right]
\end{aligned} \tag{3.54}$$

This equation is also described as

$$i\hbar\sum_n \dot{C}_n(t)U_n(\boldsymbol{x})\exp\left[-\frac{iE_n^0 t}{\hbar}\right] = \lambda\sum_n C_n V U_n(\boldsymbol{x})\exp\left[-\frac{iE_n^0 t}{\hbar}\right] \tag{3.55}$$

Let us now multiply the equation by $U_m^*(\boldsymbol{x})\exp\left[-iE_m^0 t/\hbar\right]$ and integrate the equation over all $\boldsymbol{x}$ values. Using the normalization and orthogonality of U_n, we obtain

$$i\hbar\dot{C}_m = \lambda\sum_n C_n\exp\left[\frac{i(E_m^0 - E_n^0)t}{\hbar}\right]\cdot V_{mn} \tag{3.56}$$

Where

$$V_{mn} = \int U_m^*(\boldsymbol{x})V(\boldsymbol{x},\boldsymbol{p},t)U_n(\boldsymbol{x})\mathrm{d}\boldsymbol{x} \tag{3.57}$$

and $\mathrm{d}\boldsymbol{x}$ corresponds to the volume element $\mathrm{d}x\mathrm{d}y\mathrm{d}z$ in the Cartesian coordinates. V_{mn} is the (mn)-th matrix element of V in the representation where H_0 is diagonal, and we need to note that V_{mn} is generally a function of time.

Equation (3.56) is generally an infinite set of linear equations that define each $\dot{C}_m$ in terms of all the C_n. The exact form of the solution depends on the value of each V_{mn} and the initial values of each C_n. On the other hand, V_{mn} is determined by the form of the perturbation potential and by the eigenfunctions U_n of Hamiltonian in the unperturbed system. Thus, the time variation of C_n depends on both the form of the perturbation term and the nature of unperturbed system at the starting point.

Now, our task is to find approximately how the C_n changes as a result of the perturbation potential. The approximation method we adopted is based on the fact that time variation of C_n's is proportional to λ, as we can see from Eq. (3.56). Now, let us further assume that all C_n but C_s are zero. This C_s may be taken as one except for the meaningless arbitrary phase factors. As $\dot{C}_n$ is small, we can say that, at least in some period of time after $t = t_0$ (the length of the period is determined by λ), all of the C_n are small and proportional to λ, while C_s is still almost one. Thus, to the first approximation, $\dot{C}_m$ can be solved by assigning $C_n = 0$ and $C_s = 1$ to the right-hand side of Eq. (3.53) when $m \neq s$.

$$i\hbar\dot{C}_m = \lambda\exp\left[\frac{i(E_m^0 - E_s^0)t}{\hbar}\right]\cdot V_{ms}(t)(m \neq s) \tag{3.58}$$

This equation is a good approximation until the C_m-terms calculated from this become large.

By integrating Eq. (3.58), we obtain

$$C_m = -\frac{i}{\hbar}\lambda\int_{t_0}^{t}\exp\left[\frac{i(E_m^0 - E_s^0)t}{\hbar}\right]\cdot V_{ms}(t)\mathrm{d}t \tag{3.59}$$

It is also interesting to calculate the first approximation to C_s, i.e., to the coefficients of eigenfunctions, which we used in the starting condition. For the first step, we assume $m = s$ in Eq. (3.56) and obtain

$$i\hbar\dot{C}_s = \lambda V_{ss}(t)C_s + \lambda \sum_{n\neq s} \exp\left[\frac{i(E_s^0 - E_n^0)t}{\hbar}\right] \cdot V_{sn}(t)C_n \tag{3.60}$$

Note that in the case of $n \neq s$, the sum of the right-hand sides of C_n is proportional to λ. This fact leads us to the conclusion that the term can be ignored to the first approximation, because C_n is proportional to λ^2. Thus, we obtain

$$i\hbar\dot{C}_s \cong \lambda V_{ss}(t)C_s \tag{3.61}$$

Since Eq. (3.61) can be integrated with time, a new formula is derived by the integration.

$$C_s \cong \exp\left[-\frac{i\lambda}{\hbar}\int_{t_0}^{t} V_{ss}(t)\mathrm{d}t\right] \tag{3.62}$$

The form of Eq. (3.59) can be rewritten as Eq. (3.62), if V_{ss} is not a function of time, then

$$C_s \cong \exp\left[-\frac{i\lambda V_{ss}(t - t_0)}{\hbar}\right] \tag{3.63}$$

In this stage, we may point out two substantial aspects of the above results.

(1) The absolute value of C_s is not changed by V_{ss}, since the term containing V_{ss} is only included in the exponential function. Therefore, neither probability change, nor transition will occur.
(2) To the first approximation, the V_{ss} term only has the effect of changing the angular frequency of oscillation in the wave function by $V_{ss}/\hbar$. This effect is equivalent to changing the energy in an unperturbed system by V_{ss}. Therefore, to the first approximation, the energy is expressed by

$$E = E_s + \lambda V_{ss} \tag{3.64}$$

Since

$$\lambda V_{ss} = \lambda \int U_s^* V U_s \mathrm{d}\boldsymbol{x},$$

this equation represents the mean value of perturbed potential using wave functions of an unperturbed system.

Now, $|C_m|^2$ gives the probability that the system is found in a state such that the eigenvalue of the Hamiltonian H_0 of the unperturbed system is E_m^0. Taking the probability to zero at $t = t_0$ except the incident wave, we can conclude that $|C_m|^2$ of other eigenfunction gives the probability that a transition from s-th eigenstate to m-th eigenstate of H_0 occur after the time $t = t_0$. Even though C_m continuously changes at a rate determined by the wave equation and by the boundary conditions at $t = t_0$, the system actually makes discontinuous and indivisible transitions from one state to the other. For example, we observe the state of the system by removing any perturbation potential a short time after $t = t_0$ while C_m is still very small. When you repeat this experiment for many times in succession, you can expect to find that the system is always in some eigenstate of H_0. In some cases, the system will be found in the m-th state, and the number will be proportional to $|C_m|^2$, although there are overwhelmingly many cases in which the system remains in its original state. Therefore, the perturbation potential must be regarded as causing indivisible transitions to other eigenstates of H_0.

As can be seen from Eq. (3.56), the general expression of C_m is closely related to the way of time variation in V_{mn}. There are three cases, however, in which the problem is easily solved and frequently appeared in real cases. Namely,

(A) V_{mn} is suddenly introduced at the time $t = t_0$ (V_{mn} is a constant).
(B) V_{mn} oscillates in a sinusoidal way with time progress.
(C) V_{mn} is introduced very slowly with time (an adiabatic case).

Hereinafter, we are going to calculate C_m for these three cases.

Case-A: $\boldsymbol{V_{mn}}$ **is suddenly introduced (Calculation to First Order in** $\boldsymbol{\lambda}$**)**

In this model, C_m can be derived directly from Eq. (3.59) by integration, assuming the system is initially in the first s-th eigenstate. As a result, we obtain,

$$C_m = \frac{\exp\left[i\left(E_m^0 - E_s^0\right)t_0/\hbar\right]}{E_m^0 - E_s^0}\left\{1 - \exp\left[\frac{i\left(E_m^0 - E_s^0\right)(t - t_0)}{\hbar}\right]\right\} \cdot \lambda V_{ms} \quad (3.65)$$

Equation (3.65) indicates that C_m is an oscillatory function of time. Thus, the probability that the system is in the m-th eigenstate of H_0 is described by the following equation,

$$\begin{aligned} |C_m|^2 &= \frac{\lambda^2 |V_{ms}|^2 \left|\left(1 - \exp\left[i\left(E_m^0 - E_s^0\right)(t - t_0)/\hbar\right]\right)\right|^2}{\left(E_m^0 - E_s^0\right)^2} \\ &= \frac{4\lambda^2 |V_{ms}|^2}{\left(E_m^0 - E_s^0\right)^2}\sin^2\left[\frac{\left(E_m^0 - E_s^0\right)(t - t_0)}{2\hbar}\right] \end{aligned} \quad (3.66)$$

This probability oscillates with an angular frequency of $\omega = \left(E_m^0 - E_s^0\right)/\hbar$ and exhibits a maximum every time given by

$$\frac{\left(E_m^0 - E_s^0\right)(t - t_0)}{2\hbar} = \left(N + \frac{1}{2}\right)\pi\,(N : \text{integer}) \tag{3.67}$$

The maximum value is

$$|C_m|^2_{\max} = \frac{4\lambda^2 |V_{ms}|^2}{\left(E_m^0 - E_s^0\right)^2} \tag{3.68}$$

The total probability that the system transitions away from the s-th state is equal to the sum of all $|C_m|^2$'s for all m except $m = s$. This probability is expressed by

$$\boldsymbol{P} = \sum_{s \neq m} |C_m|^2 = 4\lambda^2 \sum_{m \neq s} \frac{|V_{ms}|^2}{\left(E_m^0 - E_s^0\right)^2} \sin^2\left[\frac{\left(E_m^0 - E_s^0\right)(t - t_0)}{2\hbar}\right] \tag{3.69}$$

To accept the discussion based on the perturbation theory with approximation used in this case as a reasonable conclusion, it is necessary that $\boldsymbol{P}$ is small compared to unity. If this requirement is satisfied, C_m will be small when $m \neq s$ and C_s will be kept nearly around unity.

$$\sin^2\left[\frac{\left(E_m^0 - E_s^0\right)(t - t_0)}{2\hbar}\right] \leq 1 \tag{3.70}$$

Under Eq. (3.70), we can accordingly expect

$$\boldsymbol{P} \leq 4\lambda^2 \sum_{m \neq s} \frac{|V_{ms}|^2}{\left(E_m^0 - E_s^0\right)^2} \tag{3.71}$$

Thus, for the validity of the perturbation theory at any time, the sufficient condition is described by

$$4\lambda^2 \sum_{m \neq s} \frac{|V_{ms}|^2}{\left(E_m^0 - E_s^0\right)^2} \ll 1 \tag{3.72}$$

The abovementioned condition is always satisfied by making λ very small if there is no energy degeneracy, $E_m^0 = E_s^0$. Therefore, we can understand that the degeneracy of the energy level is important in the perturbation theory since Eqs. (3.66) and (3.68) cannot be used if such energy levels as $E_m^0 = E_s^0$ are present.

However, even in this case, C_m can also be derived from Eq. (3.58), and we obtain

$$C_m = -\frac{i}{\hbar}\lambda V_{ms}(t - t_0) \tag{3.73}$$

In this case, as shown in the form of the equation, C_m increases indefinitely with time. Thus, if the system is degenerated, the perturbation theory fails after a sufficiently long time.

It is worth to consider the physical meaning of the results obtained above. We first treat the case where there is no degeneracy. As we can see in Eq. (3.66), C_m increases with time, then decreases, then increases again, and so on. However, it never exceeds a certain finite value. This is because the system initially shifts to other quantum states. However, after the time τ given by

$$\tau = \frac{\hbar}{E_m^0 - E_s^0} \tag{3.74}$$

the reversal transition occurs. Thus, the smaller $E_s^0 - E_m^0$ is, the longer the available time for transition to m-th level and then the larger will be the obtained maximum value of C_m.

It should also be remembered in this relation that the total energy of the system is not H_0, but $H_0+\lambda V$, so that the description of the transition using the eigenstate of H_0 is not an exact description of the transition between definite energy levels. However, because λ is small, the contribution of the perturbation potential to the total energy is small (see Eq. (3.64)), E_s^0 and E_m^0 can also be interpreted as approximate eigenvalues of the energy.

Based on the abovementioned remarks, the following view description is possible for the system in which the perturbation potential is introduced. That is, the present view describes such a system kept in a state that fluctuates continuously as a superposition of one eigenstate of the non-perturbed Hamiltonian H_0 with another eigenstate, i.e., in an uncertain state of energy. In other words, when perturbation is extensively introduced, the system begins to make transitions toward any possible energy levels and becomes uncertain. If this system stays for an infinite time in one of the eigenstates of H_0 corresponding to the energy of an unperturbed system, or, in a state very different from the original energy of the unperturbed system, it would be inconsistent with the law of energy conservation. The energy conservation may hold for plenty of cases, because, as we have earlier seen, the contribution of the perturbation potential to energy is very small, but contrarily, the difference in energy $E_m^0 - E_s^0$ can generally be quite large. However, this discrepancy is avoided by the fact that the system remains in a new state for a very short period of time and that the uncertainty principle (see Eqs. (3.49) and (3.74)) makes the energy uncertain to the order of $E_m^0 - E_s^0$. Only in the case of $E_m^0 = E_s^0$, i.e., when the system is degenerated, the transition can proceed in the same direction without violating the law of energy conservation.

The above description replaces the classic general idea that a dynamical system moves along the well-defined pathway with the following new idea. Namely, under the influence of the perturbation potential, the system has a trend to transit in all "directions" (possible energy states) simultaneously. As a result, the wave function representing the state of the system is described as the sum of the contributions from many quantum states, and the system is in an uncertain state which simultaneously covers these states before reaching a stable state. The uncertainty of the energy in the state and its lifetime satisfies the uncertainty relation. On the other hand, in the case that the system is degenerate, i.e., $E_m^0 = E_s^0$, the energy is conserved. Only in this case, the transition proceeds at any rate in the same direction.

Among the transitions described above, transitions can be distinguished into real transitions and so-called virtual transitions. The former are permanent transitions in which energy is conserved, while the latter are transitions in which energy is not conserved and must return before much progress is made. The latter terminology is not appropriate, because it implies that virtual transitions have no real effect at all. The opposite is true, and they are sometimes the most important. In fact, a great number of physical processes are the results of these so-called virtual transitions.

Case-B: V_{mn} changes sinusoidally with time

We shall consider the case where the perturbation V_{mn} sinusoidally changes with time. This situation appears in many cases such as the phenomena occurring from an atom irradiated with the light of a certain angular frequency ω. We only need to consider the vector-potential $\boldsymbol{a}$ for this sort of electromagnetic phenomena. The Hamiltonian is written as follows:

$$H = \frac{\boldsymbol{p}^2}{2m} + V - \frac{e}{2mc}(\boldsymbol{a}\cdot\boldsymbol{p} + \boldsymbol{p}\cdot\boldsymbol{a}) + \frac{e^2}{2mc^2}\boldsymbol{a}^2 \tag{3.75}$$

where V is the potential created by the external force acting on atoms other than those originating from the incident electromagnetic radiation.

Since we confine our consideration to the case that small electromagnetic field is applied as a disturbance, we can ignore the second order term containing $\boldsymbol{a}^2$. We then obtain the V_{mn} by

$$\lambda V_{mn} = -\frac{e}{2mc}\int U_m^*(\boldsymbol{x})(\boldsymbol{a}\cdot\boldsymbol{p} + \boldsymbol{p}\cdot\boldsymbol{a})U_n(\boldsymbol{x})\mathrm{d}\boldsymbol{x} \tag{3.76}$$

For the light wave with a definite angular frequency, $\boldsymbol{a}$ can be written as

$$\boldsymbol{a} = \boldsymbol{G}(\boldsymbol{x})e^{-i\omega t} + \boldsymbol{G}^*(\boldsymbol{x})e^{i\omega t} \tag{3.77}$$

Note that a complex conjugate term must be added, to make the frequency $\boldsymbol{a}$ real number. Accordingly, we obtain

$$\begin{aligned}\lambda V_{mn} = -\frac{e}{2mc}\Big[&e^{-i\omega t}\int U_m^*(\boldsymbol{x})(\boldsymbol{G}\cdot\boldsymbol{p} + \boldsymbol{p}\cdot\boldsymbol{G})U_n(\boldsymbol{x})\mathrm{d}\boldsymbol{x}\\ &+ e^{i\omega t}\int U_m^*(\boldsymbol{x})\big(\boldsymbol{G}^*\cdot\boldsymbol{p} + \boldsymbol{p}\cdot\boldsymbol{G}^*\big)U_n(\boldsymbol{x})\mathrm{d}\boldsymbol{x}\Big]\end{aligned} \tag{3.78}$$

We define G_{mn} by

$$G_{mn} = -\frac{e}{mc}\int U_m^*(\boldsymbol{x})\frac{\boldsymbol{G}\cdot\boldsymbol{p}+\boldsymbol{p}\cdot\boldsymbol{G}}{2}U_n(\boldsymbol{x})\mathrm{d}\boldsymbol{x} \tag{3.79}$$

The complex conjugate can be obtained by replacing every part of the above integral with a complex conjugate form as

$$G_{mn}^* = -\frac{e}{mc}\int U_m(\boldsymbol{x})\frac{\boldsymbol{G}^*\cdot\boldsymbol{p}^*+\boldsymbol{p}^*\cdot\boldsymbol{G}^*}{2}U_n^*(\boldsymbol{x})\mathrm{d}\boldsymbol{x} \tag{3.80}$$

Writing

$$\boldsymbol{p} = \frac{\hbar}{i}\nabla$$

and noting that $\boldsymbol{G} = \boldsymbol{G}(\boldsymbol{x})$, we perform the partial integration (note that the integrated part becomes null) and obtain the next equation easily.

$$G_{mn}^* = -\frac{e}{mc}\int U_n^*(\boldsymbol{x})\frac{\boldsymbol{G}^*\cdot\boldsymbol{p}+\boldsymbol{p}\cdot\boldsymbol{G}^*}{2}U_m(\boldsymbol{x})\mathrm{d}\boldsymbol{x} \tag{3.81}$$

In other words, we can refer that $\boldsymbol{p}$ operates on U_m rather than U_n^*. With these definitions, we finally reach

$$\lambda V_{mn} = G_{mn}e^{-i\omega t} + G_{mn}^* e^{i\omega t} \tag{3.82}$$

At this stage, using Eq. (3.59), we can calculate C_m and obtain

$$\begin{aligned} C_m &= -\frac{i\lambda}{\hbar}\int_{t_0}^{t}\exp\left[\frac{i\left(E_s^0 - E_n^0\right)t}{\hbar}\right]\cdot V_{ms}(t)\mathrm{d}t \\ &= -\left[G_{ms}F(\omega) + G_{sm}^* F(-\omega)\right] \end{aligned} \tag{3.83}$$

where $F(\omega)$ satisfies

$$\begin{aligned} F(\omega) &= \exp\left[\frac{i\left(E_m^0 - E_s^0 - \hbar\omega\right)t}{\hbar}\right] \\ &\cdot\frac{\left[1 - \exp\left[-i\left(E_m^0 - E_s^0 - \hbar\omega\right)(t - t_0)/\hbar\right]\right]}{\left(E_m^0 - E_s^0 - \hbar\omega\right)} \end{aligned} \tag{3.84}$$

Thus, we finally obtain $|C_m|^2$, the probability that the system is in the m-th state. According to Eq. (3.83), we can write the probability as the form,

$$|C_m|^2 = \left|G_{ms}F(\omega) + G_{sm}^* F(-\omega)\right|^2 \tag{3.85}$$

We shall confine our interest to the case of transition processes over long period of time. Thus, we consider the case either $E_m^0 - E_s^0 = \hbar\omega$ or $E_m^0 - E_s^0 = -\hbar\omega$. The former case corresponds to the energy being absorbed from the perturbing field, while the latter corresponds to the energy being released to the perturbing field. Only the term containing $[F(\omega)F^*(\omega)]$ will actually be increased after a sufficient period of time. The term containing $F(-\omega)$ becomes oscillatory and produces only a small correction, which is ignored here. In this way, we obtain an approximation of the $|C_m|^2$ as,

$$|C_m|^2 \cong |G_{ms}|^2 |F(\omega)|^2 \tag{3.86}$$

Using the value of $|F(\omega)|^2$ in Eq. (3.84), we can also express the $|C_m|^2$ as,

$$|C_m|^2 = \frac{4|G_{ms}|^2 \sin^2\left[\left(E_m^0 - E_s^0 - \hbar\omega\right)(t - t_0)/\hbar\right]}{\left(E_m^0 - E_s^0 - \hbar\omega\right)^2} \tag{3.87}$$

Let us now turn to a discussion of the results obtained above. Except for the fact that $E_m^0 - E_s^0$ is replaced by $E_m^0 - E_s^0 \pm \hbar\omega$, the results are exactly what was obtained in Case-A (constant potential). The general results suggest that the fluctuation of $|C_m|^2$ is exactly the same as the case where V_{mn} is a constant, apart from $E_m^0 - E_s^0 = \pm\hbar\omega$. For the case satisfying $E_m^0 - E_s^0 = \pm\hbar\omega$, one of the terms [either $F(\omega)$ or $F(-\omega)$] provides a contribution that infinitely increases with time. This indicates that when the perturbation term in the Hamiltonian oscillates with the angular frequency $\hbar\omega = \left|E_m^0 - E_s^0\right|$, the perturbation term can cause irreducible transitions from the s-th level to the m-th level and from the m-th level to the s-th level. In other words, this means that the light wave with the angular frequency ω and the electrons can exchange energy without violating the law of energy conservation, only when the condition of $E_m^0 - E_s^0 = \pm\hbar\omega$ is satisfied. Since both signs of plus and minus appear in the equation, the electrons can either emit or absorb a quantum of energy, $\left|E_m^0 - E_s^0\right| = \hbar\omega$. This process has, of course, been experimentally verified.

These results can be easily described using the picture for transitions developed in the case of constant potentials. In other words, as in the Case-A, when a perturbing potential is introduced, the system fluctuates in all possible directions, and the system becomes uncertain. The uncertain relation is satisfied between the degree of uncertainty in the energy of the system and the lifetime of the system in the uncertainty state. On the other hand, the condition for irreducible transitions to be possible during periodic perturbations is Bohr-Einstein's frequency condition $\left|E_m^0 - E_s^0\right| = \hbar\omega$. Thus, we should note that, as in the previous case, the irreducible transitions, which are called virtual transitions, eventually have the most important physical effects.

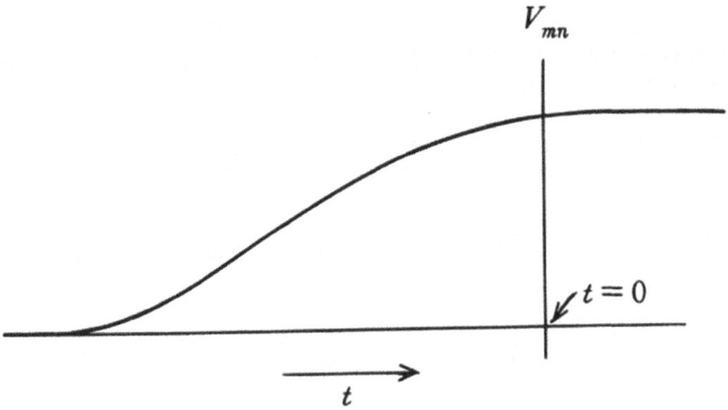

Fig. 3.3 Relation between potential and time in adiabatic condition [1]

Case-C: V_{mn} slowly changes with time (i.e., adiabatic transition)

We often find the case that the system receives a slow perturbation with time evolution. A typical variation of V_{mn} with time is shown in Fig. 3.3. V_{mn} diminishes to zero when $t \to -\infty$ and asymptotically approaches a certain constant value for the positive value of t. As we can see in Fig. 3.3, in the intermediate period of time, V_{mn} changes smoothly and slowly with time.

We shall start our consideration with Eq. (3.59) for C_m. If $t \to -\infty$, V_{mn} substantially satisfies $V_{mn} \to 0$. Thus, the error introduced by replacing t_0 with $-\infty$ in the lower limit of integration can be ignored. Accordingly, we obtain

$$C_m = -\frac{i}{\hbar}\int_{-\infty}^{t} \lambda V_{ms}(t)\exp\left[\frac{i(E_m^0 - E_s^0)t}{\hbar}\right]\mathrm{d}t \tag{3.88}$$

(N.B., for this equation to be valid, C_m must be sufficiently small.)

Partially integrating the above equation with paying attention to $V_{ms}(-\infty) = 0$, we obtain

$$C_m = -\frac{\lambda V_{ms}(t)\exp\left[i\left(E_m^0 - E_s^0\right)t/\hbar\right]}{\left(E_m^0 - E_s^0\right)} + \int_{-\infty}^{t} \frac{\exp\left[i\left(E_m^0 - E_s^0\right)t/\hbar\right]}{\left(E_m^0 - E_s^0\right)}\frac{\mathrm{d}(\lambda V_{ms})}{\mathrm{d}t}\mathrm{d}t \tag{3.89}$$

We notice that $\mathrm{d}V_{ms}/\mathrm{d}t \to 0$ as V_{ms} approaches a constant value for positive large t. Therefore, we can choose the integration limit as $\pm\infty$ when we consider times greater than $t = 0$. The error resulted from this operation should be negligible. Therefore, we obtain

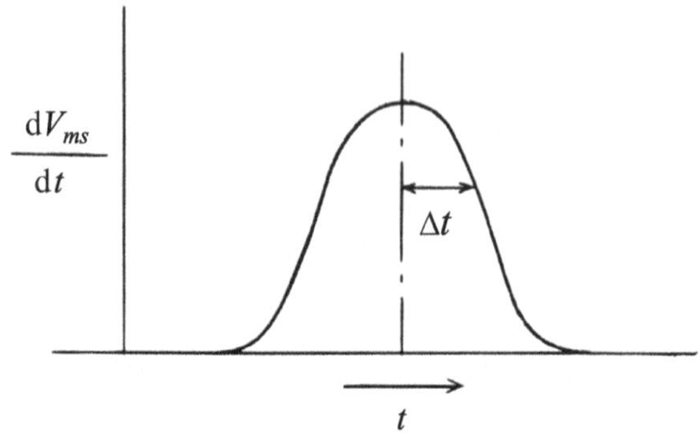

Fig. 3.4 Time variation of $\mathrm{d}V_{ms}/\mathrm{d}t$, time derivative of potential component V_{ms} [1]

$$C_m = -\frac{\lambda V_{ms}(t)\exp\left[i\left(E_m^0 - E_s^0\right)t/\hbar\right]}{\left(E_m^0 - E_s^0\right)} + \int_{-\infty}^{\infty} \frac{\exp\left[i\left(E_m^0 - E_s^0\right)t/\hbar\right]}{\left(E_m^0 - E_s^0\right)} \frac{\mathrm{d}}{\mathrm{d}t}(\lambda V_{ms})\mathrm{d}t \tag{3.90}$$

The integral on the right-hand side is exactly proportional to Fourier components of $\mathrm{d}V_{ms}/\mathrm{d}t$, which correspond to the frequencies, $\omega_{ms} = \left(E_m^0 - E_s^0\right)/\hbar$.

As schematically drawn in Fig. 3.4, $\mathrm{d}V_{ms}/\mathrm{d}t$ changes in such a way that it starts from zero at $t \to -\infty$, reaches the maximum, and then reduces to zero at $t \to +\infty$. This behavior suggests that there will be a certain mean time with a deviation of $\Delta\, t$ in which time interval $\mathrm{d}V_{ms}/\mathrm{d}t$ takes a large value. That is to say, the Fourier components of $\mathrm{d}V_{ms}/\mathrm{d}t$ are large only in the range of $\Delta\omega \sim 1/\Delta t$. Therefore, the integral on the right side of Eq. (3.90) is negligibly small under the following condition shown in the equations below

$$\Delta\omega = \frac{\left(E_m^0 - E_s^0\right)}{\hbar} > \frac{1}{\Delta t} \tag{3.91}$$

or

$$E_m^0 - E_s^0 > \frac{\hbar}{\Delta t}$$

Thus, if $\left(E_m^0 - E_s^0\right)/\hbar > 1/\Delta t$, we can write C_m as

$$C_m \cong -\frac{\lambda V_{ms}}{E_m^0 - E_s^0}\exp\left[\frac{i\left(E_m^0 - E_s^0\right)t}{\hbar}\right] \tag{3.92}$$

The physical meaning of this equation is straightforward. That is, when the potential is turned on infinitely slowly, it is concluded that only terms that oscillate with the angular frequency $\left(E_m^0 - E_s^0\right)/\hbar$ appear in C_m. If we compare this result with that obtained from Eq. (3.65) for the case that the perturbation is applied abruptly, we can find in the latter case an additional term that does not oscillate with time.

The scheme stating this Case-C is very similar to the case appearing in classic mechanics: The similar behavior can be seen when the external force of harmonically varying with angular frequency ω makes an action on a harmonic oscillator having natural angular frequency of ω_0. The equation of motion in such a dynamical system can be formulated by

$$m\left(\ddot{x} + \omega_0^2 x\right) = Fe^{i\omega t} \tag{3.93}$$

The general solution of this equation is

$$x = Ae^{i\omega_0 t} + Be^{-i\omega_0 t} + \left(\frac{F}{m}\right)\frac{e^{i\omega t}}{\left(\omega_0^2 - \omega^2\right)} \tag{3.94}$$

We can make an appropriate choice of A and B in order to satisfy any particular boundary conditions, such as,

$$x = \dot{x} = 0 \quad (\text{at } t = 0) \tag{3.95}$$

Generally, A and B are not zero. Thus, there exists the so-called free oscillation with an angular frequency of ω_0. In case, however, both A and B become zero, we can obtain only forced oscillation having an angular frequency ω which is equal to that of the forcing term.

It seems interesting to inquire how such purely forced oscillations are brought about. One way is to introduce a forcing term that works very slowly (or adiabatically, as you may call) compared to the period of the oscillation. Let us imagine the case in which the amplitude of the forcing term increases very slowly, and then, we can show that only the forced oscillation, i.e., $A = B = 0$, is obtained in the limit of infinitely slow process for making F to attain its final value. Similarly, we can regard Eq. (3.92) for C_m as the formula to determine the rate of oscillation for each one of C_m's. The term including $V_{ms}\exp\left[i\left(E_m^0 - E_s^0\right)t/\hbar\right]$ in Eq. (3.89) works as a forcing term to make the system oscillate C_m with an angular frequency $({E_m}^0 - {E_s}^0)/\hbar$. When V_{ms} increases very slowly from zero (i.e., in an adiabatic way), C_m only oscillates with the frequency of forced oscillation. However, if V_{ms} changes quite significantly in a time to the order of $\hbar/\ ({E_m}^0 - {E_s}^0)$, the free oscillation term appears in the C_m. In this case, the frequency of the free oscillation may be zero, and this type of term is just a constant. The ignorance of free oscillation can be justified by the following consideration. When there is no forcing term in Eq. (3.58), this equation becomes $i\hbar\ \dot{C}_m = 0$. Therefore, the natural frequency in this case must be regarded as zero.

Equation (3.86) is a fundamental form describing how the way of switching of the potential regulates the values of C_m. If $(\mathrm{d}V/\mathrm{d}t)_{ms}$ has Fourier components corresponding to an angular frequency $\left(E_m^0 - E_s^0\right)/\hbar$, a large constant term will be added to C_m.

This equation can also be used to describe the behavior of C_m in case of a sudden application of perturbations. To do this, we assume that V_{ms} is zero until $t = t_0$ and then it has a nonzero constant value after the time t_0. Thus, V_{ms} is a constant multiple of step function, $cS(t - t_0)$, where the normalized step function $S(t - t_0)$ is 0 for $t < t_0$ and 1 for $t > t_0$, and c is a constant. The derivative of this step function is shown to be a δ-function. Thus, the integral term in Eq. (3.89) is

$$\frac{\lambda V_{ms} \exp\left[i\left(E_m^0 - E_s^0\right)t_0/\hbar\right]}{E_m^0 - E_s^0}$$

and then, C_m becomes

$$C_m = \frac{\lambda V_{ms}}{E_m^0 - E_s^0}\left\{\exp\left[\frac{i\left(E_m^0 - E_s^0\right)t_0}{\hbar}\right] - \exp\left[\frac{i\left(E_m^0 - E_s^0\right)t}{\hbar}\right]\right\} \tag{3.96}$$

Comparing Eq. (3.96) with Eq. (3.65), we found that both equations are identical.

In the end, the above consideration leads us to two important conclusions. Based on Eq. (3.52), the first conclusion is that the complete wave function can be written as

$$\Psi = C_s U_s \exp\left(-\frac{iE_s^0 t}{\hbar}\right) + \sum_{n \neq s} C_n U_n \exp\left(-\frac{iE_n^0 t}{\hbar}\right) \tag{3.97}$$

On the other hand, according to Eq. (3.62), the C_s is expressed by

$$C_s \cong \exp\left[-i\int_{-\infty}^{t} \lambda V_{ss} \mathrm{d}t/\hbar\right] \tag{3.98}$$

Since V_{ss} is constant for $t > 0$, we get the next equation,

$$\int_{-\infty}^{t} \frac{\lambda V_{ss}}{\hbar}\mathrm{d}t = \int_{-\infty}^{0} \frac{\lambda V_{ss}}{\hbar}\mathrm{d}t + \int_{0}^{t} \frac{\lambda V_{ss}}{\hbar}\mathrm{d}t = \text{constant} + \frac{\lambda V_{ss} t}{\hbar} \tag{3.99}$$

Therefore, the C_s is expressed as

$$C_s = C_{s0} \exp\left(-\frac{i\lambda V_{ss} t}{\hbar}\right) \tag{3.100}$$

where $C_{s0} = e^{i\phi}$ and ϕ is a constant phase factor. Since the phase factor has no specific physical meaning, it can accordingly be included in the definition of U_s.

In the adiabatic case, the C_n is calculated from Eq. (3.92), and we obtain

$$\Psi = U_s \exp\left[-\frac{i\left(E_s^0 + \lambda V_{ss}\right)t}{\hbar}\right] + \sum_{n \neq s} \frac{\lambda V_{ns} U_n}{E_s^0 - E_n^0} \exp\left[-\frac{i E_s^0 t}{\hbar}\right] \tag{3.101}$$

The sum of the second term with $n \neq s$ is proportional to λ, so that the error by multiplying $\exp(-i\lambda V_{ss} t/\hbar)$ is in the order of magnitude as λ^2. Therefore, to the first approximation, we obtain

$$\begin{aligned} \Psi &= \left[U_s + \lambda \sum_{n \neq s} \frac{V_{ns} U_n(\boldsymbol{x})}{E_s^0 - E_n^0}\right] \cdot \exp\left[-\frac{i\left(E_s^0 + \lambda V_{ss}\right)t}{\hbar}\right] \\ &= f(\boldsymbol{x}) \exp\left[-\frac{i\left(E_s^0 + \lambda V_{ss}\right)t}{\hbar}\right] \end{aligned} \tag{3.102}$$

The second important conclusion drawn from Eq. (3.92) is

$$|C_m|^2 = \frac{\lambda^2 |V_{ms}|^2}{\left(E_m^0 - E_s^0\right)^2} \tag{3.103}$$

Both first and second consequences contain very interesting features. The first consequence (Eq. (3.102)) shows that the whole wave function oscillates with an angular frequency $\left(E_s^0 + \lambda V_{ss}\right)/\hbar$ (to the first approximation in λ). The system is therefore in a stationary state, and all probabilities remain constant over time. For example, the probability that this system is found in the m-th state is given by Eq. (3.102). However, we should note that this consideration only holds for the case where the perturbation is introduced very slowly. If perturbation is suddenly applied in the system, Ψ would no more take the form of $f(\boldsymbol{x})\exp\left[-i\left(E_s^0 + \lambda V_{ss}\right)t/\hbar\right]$, but the oscillation with various other frequencies would be brought about. Thus, the system is not in a stationary state, and the various probabilities will fluctuate with time. This feature is revealed in Eq. (3.66) describing the case in which the perturbation is suddenly applied at $t = t_0$.

How should we draw a picture of the origin of a stationary state? For this purpose, we can use the view that we have already created in the Case-A with a sudden introduction of a constant potential. We regard that the perturbation would tend to induce transitions invariably from the original unperturbed eigenstate with energy H_0 to other eigenstates. However, if the perturbation is slowly introduced, the dynamical system remains in a state in which the transition to another state and the transition back to the original state are balanced, so that the probability that one electron is found in another state remains constant. This probability, however, increases slowly so long as the perturbation continues to work on the system. If the perturbation is suddenly applied, the balance would not be kept, and therefore, the probability of the system transitioning to some other state would fluctuate drastically. It is similar to such a case that the free oscillation appears when the harmonic oscillator is excited abruptly. However, it should be pointed out here that this picture for describing the fluctuation is restricted in its validity. This is because of the possibility of interference between functions $U_n(x)$ different with each other. In order to take this effect into account, it is necessary to consider that the system fluctuates over all possible states at the same time and thus covers all states simultaneously.

Now, we turn to the reason why the law of energy conservation is still consistent with the existence of a certain probability that a particle will be found in the m-th state with an energy different from its original value. The clue of the reason is involved in the fact that the energy is now a sum $(H_0 + \lambda V)$. Since the wave function receiving the effect of a perturbation potential contains eigenfunctions of H_0 with a small coefficient C_n, the values of H_0 slightly fluctuate. However, the total energy is exactly $\hbar$ times the frequency ω of the wave function, i.e.,

$$E = E_s^0 + \lambda V_{ss} \tag{3.104}$$

Thereby, the spatial part of the wave function of the system is approximately expressed by

$$f(\boldsymbol{x}) = U_s(\boldsymbol{x}) + \lambda \sum_{n \neq s} \frac{U_n(\boldsymbol{x}) V_{ns}}{E_s^0 - E_n^0} \tag{3.105}$$

Thus, $H_0 + \lambda V$ has a definite value, despite that H_0 has no definite value (this explanation is just valid only for the first order approximation of λ). Thus, the function $f(x)$ is just the first approximate eigenfunction corresponding to the first approximate eigenvalue $E_s^0 + \lambda V_{ss}$ of the operator $H_0 + \lambda V$. Accordingly, a new stationary state is created due to the perturbation, if the effect is introduced very slowly.

Note that these conclusions do not hold for the case where the degeneracy is present. If any of the levels are degenerated, it is not possible to satisfy the condition of $t > \hbar/(E_s^0 - E_n^0)$ no matter how slowly the perturbation is introduced. Therefore, for degenerate levels, the perturbation theory is out of application, as we have seen, in an indefinite length of time. This implies that the adiabatic condition cannot be applied in this case, so that the conventional treatment becomes unusable.

We shall close the quantitative examinations and discussion for the change in the state of a system when a perturbation potential is introduced. From the above treatments, we can draw the conclusions that when the perturbation potential is suddenly introduced or when the perturbation varies with time, the system is in an uncertain transition state that simultaneously covers various states with different energies and that the degree of uncertainty in energy and its lifetime satisfy the uncertainty relation $\Delta t \cdot \Delta E \cong \hbar$. In the case that a perturbation is adiabatically introduced, we can regard the system as being in a stationary state for a new operator $H_0 + \lambda V$. These conclusions are valid for the first approximation in the absence of degeneracy.

3.1.3 Transition State: Characteristics and Related Problems

The transition state, which appears in various processes of phase transformations, chemical reactions, or other changes of substances, is a particularly interesting issue, and for this reason, the details of the states have been concerned as important research subjects of over the years. The transition state has been considered based on various concepts depending on the characteristics of the phenomena. For example, it is treated as nucleation in phase transformation, and as activated state or activated complex in chemical reactions. In the following, we will consider some of the problems in the treatment, based on the discussions in the previous sections, since the discussion often depends on how the transition states are treated in various phenomena and the underlying ideas.

First of all, it has been accepted that only particles with an energy equivalent to the potential peak can get over the peak in the transition state to undergo phase transformation or chemical reactions. The reason for the existence of an incubation period in phase transformation and the difficulty for chemical reactions to proceed are due to the hypothesis that this potential hill is not easily overcome.

It seems interesting to compare the classical mechanical picture of such a transition state with the results of the quantum mechanical calculation obtained in the previous section. The calculation in the previous section shows that there was indeed a time delay in passing through the peak of the potential. However, this cannot be interpreted in terms of classical mechanics. In the calculations of the previous section, it was assumed that particles have the energy greater than the height of the potential peak from the beginning.

Therefore, from the classical point of view, there should be no real potential peaks, and every particle should be able to pass through the potential barrier with no exceptions. However, taking the particle–wave duality of matter in microscopic sense into account, even without a potential peak, or conversely, even in a valley of the well-shaped potential, a particle cannot pass through the potential because the reflection occurs when the potential suddenly changes. The transmittance reaches 1 when resonance occurs. Under these conditions, all traveling particles pass through the potential barrier, but this is not because the particle received enough activation energy.

The substantial matter is whether the energy of the particle is equal to the resonance energy or not, and never because the energy of particle is not enough to overcome the potential peak. The time delay, which had been thought to correspond to the existence of an incubation period or the time required for a chemical reaction to proceed, is due to the backward reflection and forward traveling of the particles within the walls of the potential barrier. Thus, in the description of the transition state, it is inappropriate to be disposed in a classical mechanical way that a particle passes through the potential peak. Thus, the quantum mechanical effect of the duality of matter must also be taken into account.

The second issue is the assumption of definite energy in the transition state. As is evident, for example, from the fact that the energy of an activated complex in a chemical reaction is approximately estimated by solving the eigenvalue equation with help of the variational method, the definiteness of the internal energy for the transition state is assumed. Thus, the theoretical chemistry conventionally approached to the transition state as a definite one, not only in the stationary (or, equilibrium) but also in the time-dependent circumstances. However, as shown in the previous section, the transition state is uncertain in energy and satisfies the uncertainty relation, $\Delta t \cdot \Delta E \cong \hbar$. In other words, the energy is not single-valued /definite, but indefinite with a width of uncertainty. In conventional theoretical treatments, the uncertainty of energy in the transition state is neglected. If the degree of uncertainty in the transition state were small enough to be ignored, such a treatment would be justified.

However, as can be seen from the discussions in the previous section or the examples in the next section, the width of the uncertainty in the transition state is the same order of magnitude as the so-called transition energy, activation energy, or reaction energy. Thus, the existence of the uncertainty relation in the transition state is justified theoretically and experimentally with a high degree of basis.

Therefore, it is not permissible to ignore the uncertainty as used to be in the conventional treatment in theoretical chemistry. Moreover, if such notion of uncertainty is accepted, the conventional concept of activation energy, which is based on the assumption of definite energy, cannot be applied. As for the question whether it is certain or uncertain, not only the transition state is uncertain, but the state before the progress of reaction is also uncertain. However, the degree of uncertainty in the prereaction state is relatively smaller than that in the transition state.

3.2 Adoption of New Principle; Uncertainty Relation

The discussion in the former chapter has revealed that the most serious shortcomings in the conventional kinetics of phase transformation and chemical reactions are caused ultimately by ignorance of the uncertainty of states. However, not only the conventional difficulties will disappear, but also the uncertainty relation will serve as the fundamental law that gives directionality for change, if we take the duality of matter and the uncertainty states arising from it into account.

3.2.1 Transition State and Uncertainty Relation

From the discussion in the previous chapter, the value Δt is referred to the duration (lifetime) of a state (e.g., a transition state) and ΔE to the uncertainty width in energy required to realize the state in the uncertainty relation $\Delta t \cdot \Delta E \cong \hbar$. Table 3.1 shows the examples which may confirm that the uncertainty relation is justified between time and energy. This table contains eight examples: the interaction between nucleons and

Table 3.1 Phenomena which satisfy $\Delta t \cdot \Delta E \cong \hbar = 6.58 \times 10^{-16}$ eV s [3]

No	Phenomenon	Δt (s)	ΔE (eV)	Remarks			
				$\Delta t \cong a/v$	ΔE	Example	Ref
1	Nuclear fission	~10^{-23}	~ 10^{8}	$\Delta t = 10^{-13}$ cm/10^{10} cm/s = 10^{-23} s	$\Delta E \cong \hbar/\Delta t = 10^{8}$ eV,(a: nuclear diameter)	^{235}U	[4, 5]
2	α-, β-decay	~10^{-21}	~ 10^{6}	$\Delta t = 10^{-13}$ cm/10^{8} cm/s = 10^{-21} s	$\Delta E \cong \hbar/\Delta t = 10^{6}$ eV	^{226}Ra ^{234}Th	[4]
3	e^+,e^- annihilation	~10^{-21}	~ 10^{6}	$\Delta t = 10^{-11}$ cm/10^{10} cm/s = 10^{-21} s	$\Delta E \cong \hbar/\Delta t = 10^{6}$ eV $= m_{e\pm}C^2$	–	[5]
4	Electromagnetic radiation	~10^{-15}	~ 1	$\Delta t = 10^{-8}$ cm/10^{7} cm/s = 10^{-15} s	$\Delta E \cong \hbar/\Delta t = 1$ eV	–	[1]
5	Chemical reaction	~10^{-15}	~ 1	$\Delta t = 10^{-8}$ cm/$10^{7\sim 8}$ cm/s = $10^{-15\sim -16}$ s	$\Delta E \cong \hbar/\Delta t = 1$ eV	–	[1, 6]
6	Martensitic transformation	~10^{-13}	~10^{-2}	$\Delta t = 10^{-8}$ cm/10^{5} cm/s = 10^{-13} s	$\Delta E \cong \hbar/\Delta t = 0.01$ eV (metamorphosis heat)	–	[7]
7	Van der Waals force	~10^{-12}	~10^{-3}	$\Delta t = 10^{-8}$ cm/10^{4} cm/s = 10^{-12} s	$\Delta E \cong \hbar/\Delta t = 10^{-3}$ eV	–	[1]
8	Mössbauer effect	~10^{-7}	~10^{-8}	–	–	^{57}Fe	–

Annotation

1. Time–energy uncertainty principle $\Delta t \cdot \Delta E \cong \hbar$, Δt means duration time of the state, ΔE means fluctuation width of energy
2. $\Delta t = a/v$, v means velocity of particle which is related with the phenomenon, a means size of an area in which the particle can move

the nuclear force field, the interaction between charged particles and the Coulomb potential, the interaction between charged particles and the electromagnetic field, etc. The width of energy ΔE spreads from 10^{8} to 10^{-8} eV, and we notice that the examples cover a wide range of phenomena from very strong interactions to very weak interactions.

For the first example of this table, ΔE is of course the energy generated by the fission of ^{235}U, approximately 10^8 eV, and Δt is the time of fission, 10^{-23} s. The product of these two values is approximately equal to $\hbar = 6.58 \times 10^{-16}$ eV·s. Here, the reaction time of fission Δt is explained as follows. According to the nuclear theory, the nuclear force acting between nucleons is explained by the exchange of π mesons (quanta of the nuclear force field). Thus, according to this idea, the reaction time Δt is roughly estimated as the time it takes for π-mesons with velocity $v = 10^{10}$ cm /s to fly through a nucleus of size $a = 10^{-13}$ cm. Thus, $\Delta t \times a/v = 10^{-23}$ s. When Δt is obtained, the energy is determined as $\Delta E \cong \hbar/\Delta t \cong 10^8$ eV. Alternatively, we can assign the mass of π-meson as m using the relation $\Delta E = mc^2$ in the theory of relativity, and then, we get $\Delta E = m_{\pi-meson}c^2 \cong 10^8$ eV. As noted in remarks of this table, other examples may be similarly explained.

The conclusions that can be drawn from this table are, first, that the physical meaning of the uncertainty relation $\Delta t \cdot \Delta E \cong \hbar$ where Δt is the duration (lifetime) of a state and ΔE is the uncertainty width in energy required to realize that state, is experimentally confirmed to be correct, and second, that a very wide range of phenomena satisfy the uncertainty relation and, therefore, this relation can be applied to various phenomena with a high reliability.

3.2.2 Application of Uncertainty Principle $\Delta t \cdot \Delta E \cong \hbar$ to Our Reaction Kinetics

As pointed out in Chap. 2, the crucial difficulty in the conventional theories of reaction kinetics or phase transformation kinetics is absence of the fundamental laws that give the direction of change. Therefore, in order to develop a new reaction kinetics, we looked for the fundamental law that gives the direction of change. As already pointed out in Chap. 1, there are several laws related to irreversibility in some sense: the discontinuous, non-causal, irreversible change of the wave function in quantum mechanical observation, the one-way change of the Boltzmann's H-function, and the law of increasing entropy in thermodynamics. Among them, the irreversible change of the wave function does not seem to be very useful as a basic law that gives the directionality of change in reaction kinetics; the irreversible change of the H-function and the entropy seem to be useful if you think about the directionality for a moment, but they do not include the concept of time.

Since reaction kinetics is concerned with the time variation of a physical quantity (e.g., the amount of a substance), anything that does not include the concept of time is not useful as the fundamental law we are looking for. In the end, we had to abandon them for the basic principle. Thus, we have no choice except for seeking an entirely new concept. We turned to find a physical rule really containing the concepts of time and directionality for change. In fact, we did not have to search so hard for it; we have already kept it in our hands. It is the time–energy uncertainty relation $\Delta t \cdot \Delta E \cong \hbar$. This relation means that the lifetime of a particle in a certain state is determined by the fluctuation width of its energy. Therefore, this equation is the very suitable and fundamental law that we need.

Now, we have found the basic law that gives the directionality for change. It is easy to see that the reaction kinetics based on this uncertainty relation will provide us with a much more satisfactory perspective than the conventional reaction kinetics. **First**, the uncertainty relation includes the concept of time and the directionality of change, as already mentioned. **Second**, the uncertainty relation is expressed in terms of time and energy, which are very general and fundamental concepts, suggesting that the relation has a wide range of applications and thus qualifies as a fundamental law in reaction kinetics. **Third**, in conventional reaction kinetics, equilibrium is assumed between the reactants and the activated complex. However, this assumption of equilibrium along the direction for the procession of change is extremely unreasonable when discussing the changes. Based on the uncertainty relation, all requiring parameters are the fluctuation width of energy and lifetime. Thus, an unreasonable assumption of equilibrium along the direction of change is not necessary. These three points are the general advantages of the reaction kinetics based on the uncertainty relation.

In the end, we shall briefly touch on the way to construct our reaction kinetics based on this uncertainty relation. The uncertainty relation means that if we know the width of the energy fluctuation in a system, we can determine the lifetime of the system. Therefore, if a reaction proceeds by some physical factors, what we need to do for understanding the mechanism or the rate of the reaction is to calculate the width of energy fluctuation caused by the physical factors. For example, if the reaction proceeds by increasing the temperature, we shall explain the variation of the reaction rate with elevation in temperature by (1) calculating the width of energy fluctuation caused by the elevation in temperature and then (2) being able to know the lifetime of the reactants. Such energy fluctuations are not caused exclusively by the increase in temperature. But mechanical, optical, electromagnetic, or any other external actions that produce energy fluctuations can be also effective. Thus, what we need to do is to evaluate the width of energy fluctuation due to these various actions.

References

1. D. Bohm, Chapters 8, 11, 18, in *Quantum Theory* (Prentice-Hall Inc., 1951)
2. A. Messiah, *Mécanique Quantique*, (Dunod, 1959), (English version, *Quantum Mechanics*, Vol. 1, Chap.6, 8, (North-Holland Publishing, Amsterdam/Oxford, 1975))
3. S. Yamamoto, The time energy uncertainty principle and thermal activation. Z. Phys. Chem. **290**, 17–32 (1989). https://doi.org/10.1515/zpch-1989-27003
4. R. E. Peierls, *The Law of Nature* (George Allen & Unwin, 1955)

5. В.И. Рыдник, *Что Такое Квантовая Мханика* (V. I. Rydnik,*what is quantum mechanics?*), translated into Japanese by H. Toyota, (Tokyo-Tosho, Tokyo, 1965)
6. Y. Yoshizawa, *What is element? [in Japanese]* (Kodansha, Tokyo, 1975)
7. J.W. Christian, *The Theory of Transformation in Metals and Alloys* (Pergamon Press, 1965)
8. H. Tanaka, *The True Face of Quantum* (Otsuki Shoten, Tokyo, 1976)

Further Readings

L.I. Schiff, *Quantum Mechanics* (McGraw-Hill Book, 1968)
S. Tomonaga, *Quantum Mechanics I, II* (Misuzu Shobo, Tokyo, 1969)
Л. Д. Ландау и Е. М. Лифшниц: *Квантовая Механика*, translated into Japanese by K. Sasaki, S. Koumura, (Tokyo-Tosho, Tokyo,1971), (English version, L. D. Landau and E. M. Lifshitz: *Quantum Mechanics*, 3nd edn. (Pergamon Press, 1981)
W.C. Price, S.S. Chissik, *The Uncertainity Principle and Foundations of Quantum Mechanics* (Wiley, A fifty years' survey, 1977)

Chapter 4
Mathematical Formulation of Our Theory

Abstract In this chapter, we will derive mathematically the Arrhenius equation. From the duality of the wave and particle nature of matter, we would derive simultaneously two uncertainty relations of time–energy and position momentum while maintaining symmetry as much as possible. Next, referring ΔE be the energy width superimposed to create the wave packet and Δt be the time width when the wave intensity is nonzero, and we show that the relation $\Delta t \cdot \Delta E \cong \hbar$ holds between them. Thermal activation which is expressed by the Arrhenius equation is interpreted as an increase in the fluctuation width of the energy due to the increase in temperature, and therefore, the lifetime is shortened from the uncertainty relation. The fluctuation width due to thermal activation was calculated using the density matrix. The Arrhenius equation $\kappa = A\exp(-E/kT)$ is an empirical expression, and it is well applicable to actual measurements. So-called activation energy E was turned out to be quantized energy-level spacing. Frequency factor A was related to E, which means specific rate constant κ was expressed by quantized factor.

Keywords Arrhenius equation · Thermal activation · Bonding energy · Wave packet · Harmonic oscillator · Density matrix

We have already stated in Chap. 3 that the transition state is a quantum mechanical uncertainty state, and that its quantitative expression is the uncertainty relation, $\Delta t \cdot \Delta E \cong \hbar$. By using it as a basic law that gives a directionality for change, we obtained a physical prospect that a new reaction kinetics could be constructed. In this chapter, we will carry out mathematical formulation of a new reaction kinetics according to this physical perspective. Let us begin with mathematical formulation of the uncertainty relation $\Delta t \cdot \Delta E \cong \hbar$ and calculate the width of energy fluctuation caused by various factors such as temperature or others, and then construct specific rate equations of reactions based on the results of calculation.

S. Yamamoto, *Reaction Kinetics Based on Time-Energy Uncertainty Principle*,
https://doi.org/10.1007/978-981-19-9673-3_4

4.1 Uncertainty Relation

In this section, we first mathematically formulate the uncertainty relation. What is used directly in the formulation of reaction kinetics is the time–energy uncertainty relation, but its physical meaning and the conditions for its establishment, the characteristics of mathematical treatment, and another uncertainty relation related with position and momentum will be discussed. For simplicity, let us first generally solve these two uncertainty relations simultaneously and in keeping a symmetrical way as possible. Then, a more detailed treatment of $\Delta t \cdot \Delta E = \hbar$ will be performed to help in the formulation of the reaction kinetics.

4.1.1 Duality of Matter and Two Uncertainty Relations

As is well known in classical wave theory, a wave having a certain wavelength λ and a frequency ν is represented by the following equation as a function of position x and time t.

$$\Psi(x,t) = \psi_0 \exp(i\boldsymbol{k}x - i\omega t) \tag{4.1}$$

where $\boldsymbol{k} = 2\pi / \lambda$ is a wave number and $\omega = 2\pi\nu$ is an angular frequency. The wave represented by Eq. (4.1) undulates from $x = -\infty$ to $x = +\infty$ and from $t = -\infty$ to $t = +\infty$. The waves we were actually interested in are not such waves that spread infinitely in space and time, but waves that exist in a finite space–time region. For example, to describe the motion of a pulse localized in a comparatively narrow area, it is necessary to obtain a wave restricted to a certain area of space. To do so, we must make what is known as a wave packet. Wave packet comprises a group of waves having slightly different wavelengths, with their phases and amplitudes so chosen that they interfere with each other in a constructive manner only over a small area of space, outside of which the amplitude rapidly becomes zero as a result of destructive interference. We can construct a wave packet, for example, by integrating a plane wave, $\exp[ik(x - x_0)]$, over a small range of wavelength ($\Delta\boldsymbol{k} \ll \boldsymbol{k}_0$). When drawn as a function of $(x - x_0)$, the real part of $\Psi(x)$ looks like the curve shown in Fig. 4.1.

$$\Psi(x) = \int_{k_0-\Delta k}^{k_0+\Delta k} \mathrm{d}\boldsymbol{k} \exp[i\boldsymbol{k}(x - x_0)] = \frac{2\sin\Delta\boldsymbol{k}(x - x_0)}{(x - x_0)} \exp[i\boldsymbol{k}_0(x - x_0)] \tag{4.2}$$

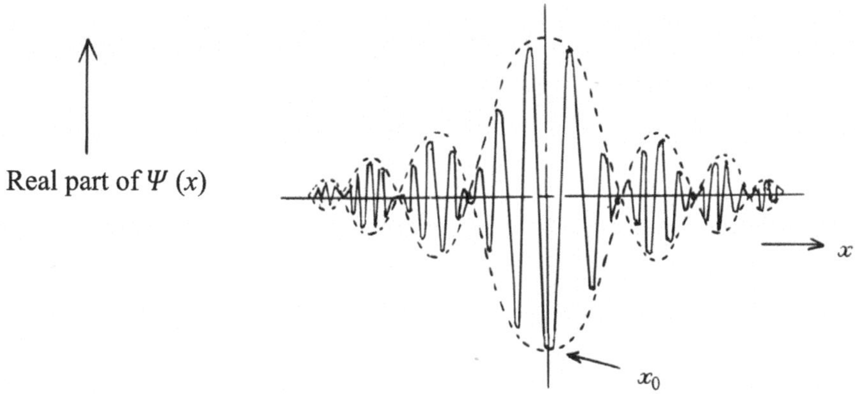

Fig. 4.1 Oscillatory variation of real part of Ψ (x) function in terms of x, showing an example of a wave packet in space [1]

This is an oscillatory function; the amplitude of oscillation reaches a maximum at $x = x_0$ and drops to zero at $x - x_0 = \pi/\Delta \boldsymbol{k}$, and then decreases rapidly. Both of the real and imaginary parts of the function $\Psi(x)$ oscillate rapidly as a function of $(x - x_0)$. Wave strength is proportional to the square of the maximum amplitude of the oscillation. If the wavelength ($\lambda = 2\pi/k_0$) is much smaller than the width of the wave packet $\Delta \boldsymbol{k}$, though this is usually the case, this maximum is very well approximated by the square of the absolute value of the complex function $\Psi(x)$. Therefore, we have

$$I \sim |\Psi(x)|^2 = \frac{4\sin^2 \Delta \boldsymbol{k}(x - x_0)}{(x - x_0)^2} \tag{4.3}$$

The strength I given by Eq. (4.3) begins to become small enough when $\Delta x = x - x_0$ is fairly greater than $1/\Delta \boldsymbol{k}$, or when $\Delta x > 1/\Delta \boldsymbol{k}$. Since $\Delta \boldsymbol{k}$ is a measure of range in wave number $\boldsymbol{k}$ to create a wave packet, the product of the width of the wave packet in $\boldsymbol{k}$-space and its width in x-space is about 1. That is, we obtain

$$\Delta x \cdot \Delta \boldsymbol{k} \cong 1 \tag{4.4}$$

This means that a wave packet with a narrow range in $\boldsymbol{k}$-space should be much wider in x-space, and vice versa. If exactly the same procedure as for the wave packet in space is applied to the wave packet in time given by

$$\Psi(t) = \int_{\omega_0-\Delta\omega}^{\omega_0+\Delta\omega} \mathrm{d}\omega \exp[-i\omega(t - t_0)] = \frac{2\sin\Delta\omega(t - t_0)}{(t - t_0)}\exp[-i\omega_0(t - t_0)] \quad (4.5)$$

We obtain the similar result,

$$\Delta t \cdot \Delta\omega \cong 1 \quad (4.6)$$

where $\Delta t = t - t_0$. The above Eqs. (4.4) and (4.6) are general properties of classical waves and are not limited to quantum theory.

Now, let us make a quantum mechanical interpretation to the quantities appearing in these equations. According to quantum mechanics, all materials exhibit the dual properties of particle and wave, and all energies take a form of quantum. Assuming that the momentum and energy of a substance as a particle are p and E and the wavelength and frequency of a substance as a wave are λ and ν, respectively, the dual (particle and wave) properties of the substance are unified as,

$$p = \frac{h}{\lambda} = \hbar\boldsymbol{k} \quad \text{(de Broglie equation)} \quad (4.7)$$

and

$$E = h\nu = \hbar\omega \text{ (Plank–Einstein equation)} \quad (4.8)$$

where the Planck constant $h = 6.626 \times 10^{-27}$ erg s $= 3.6 \times 10^{-15}$ eV s. $\hbar = h/2\pi$

By using Eqs. (4.7) and (4.8), let us interpret the above discussion based on the classical wave theory. Since the wave function of a substance with momentum p is exp $(ipx/\hbar)$ and the wave function of a substance with energy E is expressed as exp $(-iEt/\hbar)$, the wave function corresponding to Eq. (4.1) is given by

$$\Psi = \psi_0 \exp(ipx/\hbar - iEt/\hbar) \quad (4.9)$$

A wave with the momentum p and the energy E, expressed in Eq. (4.9), propagates from $x = -\infty$ to $x = +\infty$ over $t = -\infty$ to $t = +\infty$. The wave packets corresponding to Eqs. (4.2) and (4.5) are created by integrating them over small region of momentum and region of energy, respectively. The expressions corresponding to Eqs. (4.4) and (4.6) are

$$\Delta x \cdot \Delta p \cong \hbar \tag{4.10}$$

and

$$\Delta t \cdot \Delta E \cong \hbar \tag{4.11}$$

Thus, we can find that the uncertainty relation for position and momentum is derived from the description by a wave packet using de Broglie Eq. (4.7), and the uncertainty relation for time and energy is derived from the description by a wave packet using the Planck–Einstein Eq. (4.8), respectively. The following is what we learned from the above derivation. The momentum of a system that can be described by a wave packet in terms of space is not definite but has a certain spread, and similarly, the energy of a system that can be described by a wave packet in terms of time is not definite but has a certain spread. The uncertainty relations hold as long as a substance exhibits the particle–wave duality.

4.1.2 Absence of Fluctuation and Eigenvalue Equations

Fluctuations in the statistical distribution of dynamical variable A are measured by the standard deviation ΔA:

$$\begin{aligned}(\Delta A)^2 &= \langle (A - \langle A \rangle)^2 \rangle = \left\langle \left(A^2 - 2A\langle A \rangle + \langle A \rangle^2 \right) \right\rangle \\ &= \langle A^2 \rangle - \langle A \rangle^2 \qquad (\geq 0)\end{aligned} \tag{4.12}$$

A^2 is also a dynamical variable like A, and $\langle A^2 \rangle$ is its mean value. When the deviation ΔA is 0, there is no fluctuation, and we can say with certainty that A has a definite value. Its value is clearly equal to $\langle A \rangle$ [2, Chap. 5]. Let us examine what condition the equation $\Delta A = 0$ imposes on the function Ψ. If $\Delta A = 0$, then from Eq. (4.12)

$$\langle A^2 \rangle = \langle A \rangle^2 \tag{4.13}$$

The mean values of operators A and A^2 are defined by the following equation

$$\langle A\rangle = \frac{\langle\psi, A\psi\rangle}{\langle\psi, \psi\rangle} \text{ and } \langle A^2\rangle = \frac{\langle\psi, A^2\psi\rangle}{\langle\psi, \psi\rangle} \tag{4.14}$$

Therefore, the relation of (4.13) is written as

$$\langle\psi, A^2\psi\rangle\langle\psi, \psi\rangle = \langle\psi, A\psi\rangle^2 \tag{4.15}$$

The quantity $\langle\psi, A^2\psi\rangle \equiv \langle\psi, A(A\psi)\rangle$ is equal to $\langle A\psi, A\psi\rangle$ due to the Hermitian nature of the operator.

Therefore, Eq. (4.15) is

$$\langle A\psi, A\psi\rangle\langle\psi, \psi\rangle = \langle\psi, A\psi\rangle^2 \tag{4.16}$$

Here, we use Schwartz inequality:

$$\sqrt{\langle\varphi, \varphi\rangle\langle\psi, \psi\rangle} \geq |\langle\varphi, \psi\rangle| \tag{4.17}$$

Equality holds only when one of the functions φ and ψ is a constant multiple of the other. Equation (4.16) is for the case where Schwartz inequality (4.17) reduces to the equality, so the functions ψ and $A\ \psi$ are proportional. That is, when a is a constant, we obtain

$$A\psi_a = a\psi_a \tag{4.18}$$

Equation (4.18) is an eigenvalue equation. Thus, we have obtained the eigenvalue equation from the condition that there is no fluctuation ($\Delta A = 0$). Examples of this type of equation are already given by the time-independent wave Eq. (3.3). Thus, the following conclusion is reached. When the fluctuation is absent in physical quantity A, the quantity A certainly has a definite value with certainty (i.e., with a probability equal to 1), and the state of the physical system at that time is represented by the eigenfunction ψ_a of the Hermitian operator A. The value of the quantity A is the eigenvalue a corresponding to the function ψ_a.

4.1.3 Wave Packet and Uncertainty Relation $\Delta t \cdot \Delta E \cong \hbar$

In order to develop a new reaction kinetics later, let us summarize here the relationship between the wave packet and the uncertainty relation $\Delta t \cdot \Delta E \cong \hbar$. For this purpose, the function is expanded into a Fourier series or a Fourier integral as shown below.

From the mathematical analysis, it is known that the periodic function $f(t)$ can always be expanded into components corresponding to harmonic oscillations. That is, according to Fourier series theory,

$$
\begin{aligned}
f(t) &= \sum_{-\infty}^{\infty} a_n(\nu) \exp(-i2\pi n\nu t) \\
a_n(\nu) &= \frac{1}{T} \int_{-T/2}^{T/2} f(t) \exp(i2\pi n\nu t) \mathrm{d}t \quad (T : \text{period})
\end{aligned}
\tag{4.19}
$$

Similarly, for the non-periodic function $f(t)$,

$$
\begin{aligned}
f(t) &= \int_{-\infty}^{\infty} a(\nu) \exp(-i2\pi \nu t) \mathrm{d}\nu \\
a(\nu) &= \int_{-\infty}^{\infty} f(t) \exp(i2\pi \nu t) \mathrm{d}t
\end{aligned}
\tag{4.20}
$$

Mathematical expression of a function by a Fourier series or a Fourier integral physically means expanded into a spectrum. When a periodic function is expanded into a Fourier series, discrete line spectra with frequencies ν, 2ν, 3ν, ... are obtained. On the other hand, when the non-periodic function is expanded into Fourier integrals, the frequency becomes a continuous spectrum having a certain spread.

Hereafter, we will consider the various cases along Table 4.1.

Table 4.1 Wave packet and the uncertainty relation $\Delta t \cdot \Delta E \cong \hbar$

	Amplitude $\psi(t)$	Probability (square of the amplitude) $P(t) = \psi^*(t)\psi(t)$
1	$\psi(t) = \exp[-iE_n t/\hbar]\varphi_n(q)$ 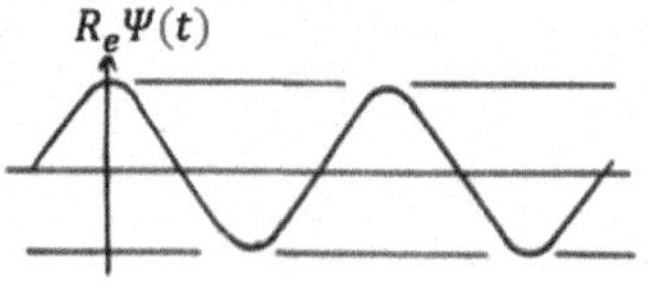	$P(t) = \psi^*(t)\psi(t) = \varphi_n(q)^2 = \text{const}$ 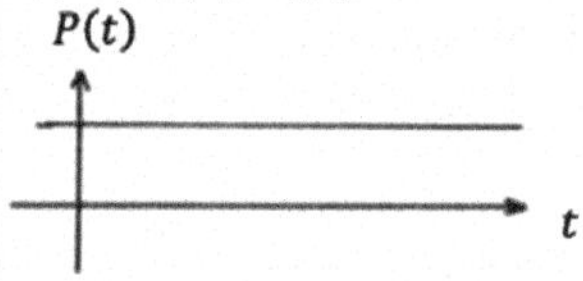
2	$\Psi(t) = c_1 \exp[-iE_1 t/\hbar]\varphi_1(q) + c_2 \exp[-iE_2 t/\hbar]\varphi_2(q)$ 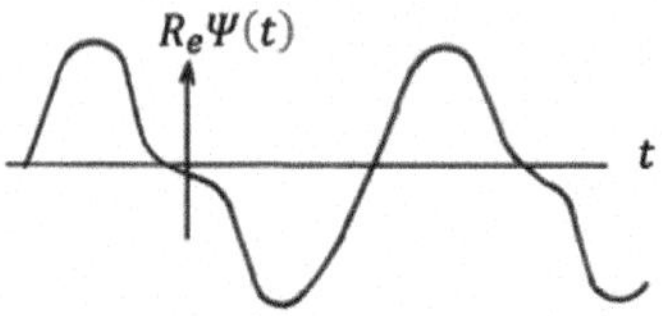	$P(t) = c_1^* c_1 \varphi_1^*(q)\varphi_1(q) + c_2^* c_2 \varphi_2^*(q)\varphi_2(q) + c_1^* c_2 \varphi_1^*(q)\varphi_2(q)\exp[-i(E_2 - E_1)t/\hbar] + c_1 c_2^* \varphi_1(q)\varphi_2^*(q)\exp[-i(E_1 - E_2)t/\hbar]$ 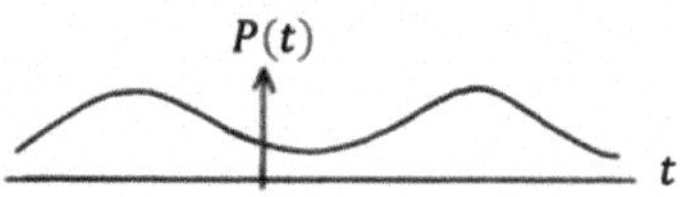
3	$\Psi(t) = \sum_n c_n \varphi_n(q)\exp[-iE_n t/\hbar]$ 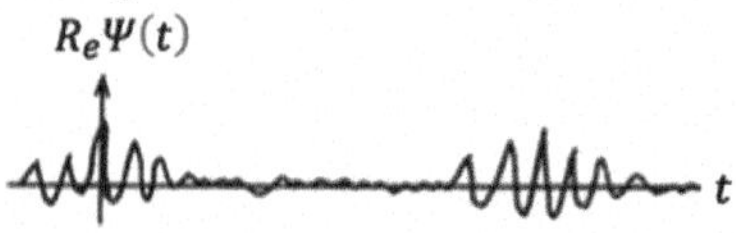	$P(t) = \Psi'\Psi = \sum_n \sum_{n'} c'_n c_{n'} \varphi_n^*(q)\varphi_{n'}(q)\exp[-i(E_n - E_n)t/\hbar]$ 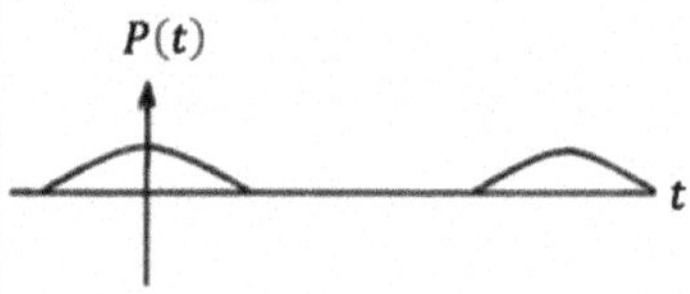
4	$\Psi(t) = \int_{-\infty}^{+\infty} a(v)\exp[-i2\pi vt]\mathrm{d}v$ 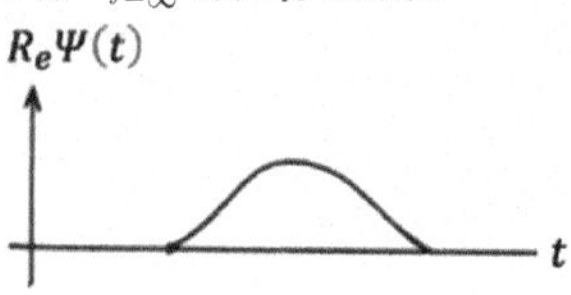	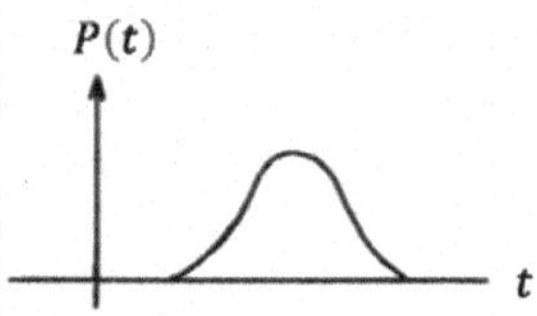
5	$\Psi(t) = \begin{cases} \varphi_0 : \text{const } 0 < t < \tau_0 \\ 0 \quad t < 0, \quad t > \tau_0 \end{cases}$ 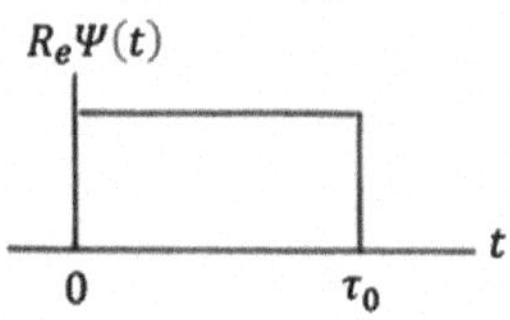	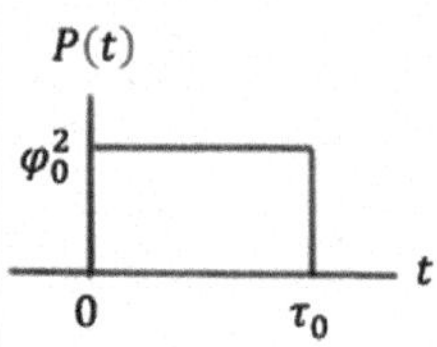

(continued)

Table 4.1 (continued)

Energy spectrum (Fourier component) $a(v) = \int_{-\infty}^{+\infty} \psi(t) \exp[i 2\pi v t] dt$	ΔF Δt	Notes
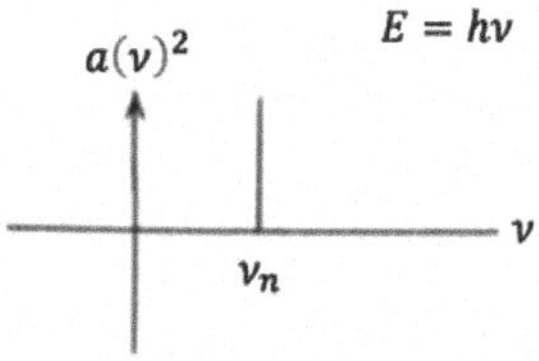	$\left.\begin{array}{l}\Delta E = 0 \\ \Delta t = \infty\end{array}\right\} \Delta t \cdot \Delta E \cong \hbar$	Solution of wave equation $\Delta E = \sqrt{\langle (E_n - \langle E_n \rangle)^2 \rangle} = 0$ ↓ $H\psi_n = E_n \psi_n$
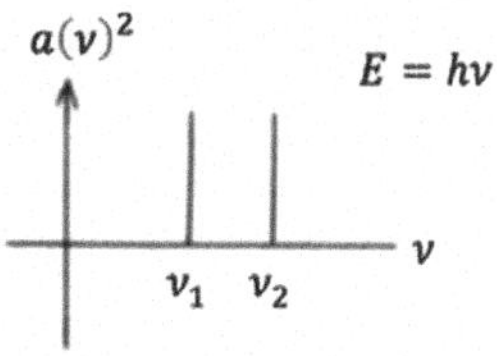	$\left.\begin{array}{l}\Delta E = E_2 - E_1 \\ \Delta t = \frac{2\pi h}{E_2 - E_1}\end{array}\right\} \Delta t \cdot \Delta E \cong \hbar$ (Period/Reaction time)	Superposition of two wave functions with energies $E1$ and E_2
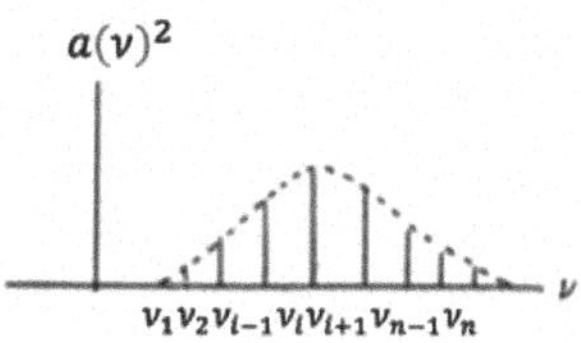	$\left.\begin{array}{l}\Delta E = E_n - E_1 \\ \Delta t = \frac{2\pi h}{E_n - E_1}\end{array}\right\} \Delta t \cdot \Delta E \cong \hbar$	Superposition of wave n functions with energies $E_1, E_{2'}, \cdots, E_x$
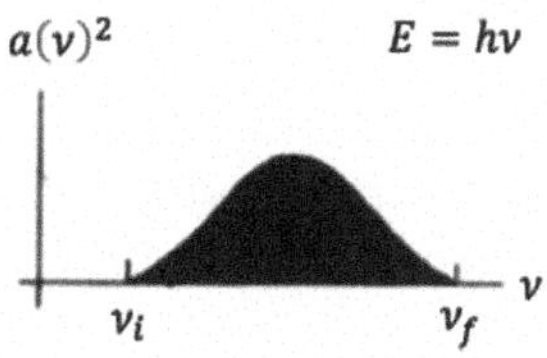	$\left.\begin{array}{l}\Delta E = E_f - E_i \\ \Delta t = t_f - t_i\end{array}\right\} \Delta t \cdot \Delta E \cong \hbar$	Wave packet (refer to 3.3.1)
$a(v)^2 = (\varphi_0 \tau_0)^2 \times \left[\frac{\sin(\pi v \tau_0)}{\pi v \tau_0}\right]^2$ 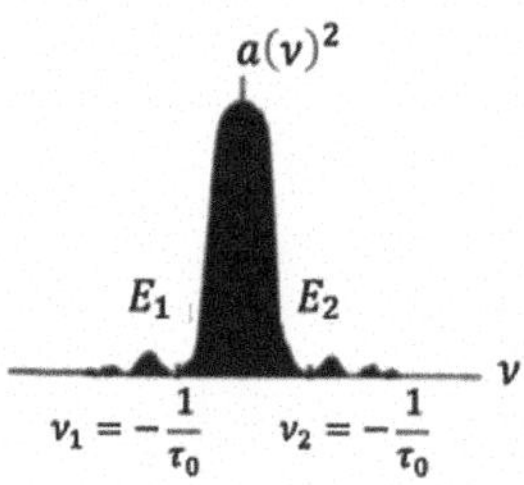	$\left.\begin{array}{l}\Delta E = E_2 - E_1 \\ \Delta t = \tau_0\end{array}\right\} \Delta t \cdot \Delta E \cong \hbar$	

1) **The first example in Table 4.1**

Let φ_n be the solution of the time-independent wave equation $H\varphi = E\varphi$ corresponding to the stationary state, and let the corresponding energy be E_n ($n = 1, 2, \ldots$). Then, the wave function is expressed in Eq. (4.21) as a function of t.

$$\Psi(t) = \varphi_n(q) \cdot \exp\left(-\frac{iE_n t}{\hbar}\right) = \varphi_n(q) \cdot \exp(-i2\pi\nu_n t) \quad (E_n = h\nu_n) \tag{4.21}$$

This Fourier component is given by using Eq. (4.19) as,

$$a_n(\nu) = \varphi_n(q) \tag{4.22}$$

From these two equations, the probability $P(t)$ of finding a particle at a certain position q is

$$P(t) = \Psi^*(t)\Psi(t) = \varphi_n(q)^*\varphi_n(q) = |\varphi_n(q)|^2 = \text{const} \tag{4.23}$$

which is constant and independent of time. This means that the particle under consideration stably remains in a given state for an indefinitely long time, and the lifetime $\Delta t = \infty$ (a stationary state). On the other hand, as is clear from the discussion in the previous section, considering the eigenvalue equation $H\varphi = E\varphi$ means that ΔE, the energy fluctuation of the system, is zero. Thus, the results shown in the first example of Table 4.1 are obtained. That is, in the eigenstate, the real part of the wave function is a cosine function of time. Therefore, the probability $P(t)$ is constant independent of time, and the energy spectrum is a single line spectrum of $E_n = h\nu_n$. In this case, $\Delta t = \infty$ and $\Delta E = 0$ in the uncertainty relation $\Delta t \cdot \Delta E \cong \hbar$.

2) **The second example**

This is the superposition of the eigenfunctions, $\exp[-iE_1t/\hbar] \cdot \varphi_1(q)$ and $\exp[-iE_2t/\hbar] \cdot \varphi_2(q)$, corresponding to two different energies E_1 and E_2. In this case, the wave function:

$$\Psi(t) = c_1\exp\left(-\frac{iE_1t}{\hbar}\right) \cdot \varphi_1(q) + c_2\exp\left(-\frac{iE_2t}{\hbar}\right) \cdot \varphi_2(q) \tag{4.24}$$

and the corresponding probability:

$$\begin{aligned} P(t) = \Psi^*(q)\Psi(q) &= c_1^*c_1\varphi_1^*(q)\varphi_1(q) + c_2^*c_2\varphi_2^*(q)\varphi_2(q) \\ &+ c_1^*c_2\varphi_1^*(q)\varphi_2(q)\exp\left[-\frac{i(E_2 - E_1)t}{\hbar}\right] \\ &+ c_1c_2^*\varphi_1(q)\varphi_2^*(q)\exp\left[-\frac{i(E_1 - E_2)t}{\hbar}\right] \end{aligned} \tag{4.25}$$

are a periodic function that changes with $2\pi\hbar/(E_2 - E_1)$ as a period. In this case, the energy spectrum is two line spectra of $E_1 = h\nu_1$ and $E_2 = h\nu_2$. In addition, $\Delta t = 2\pi\hbar/(E_2 - E_1)$ and $\Delta E = E_2 - E_1$ in $\Delta t \cdot \Delta E \cong \hbar$.

3) **The third example**

This is a case where n wave functions with different energies $E = E_1, E_2,.., E_n$. It can be considered in the same way as in the second example, and the energy spectrum becomes n line spectra, and $\Delta t = 2\pi\hbar/(E_n - E_1)$, $\Delta E = E_n - E_1$.

The above three examples are for the case where the wave function is a periodic function of time, and therefore, the spectrum is a discrete line spectrum, as described above.

Next, let us consider two cases where the wave function is a non-periodic function of time.

4) **The fourth example**

The forth is corresponding to the most common case. In this case, as is clear from the derivation of $\Delta t \cdot \Delta E \cong \hbar$ in Sect. 4.1.1 and the treatment of the wave packet in Sect. 3.1.1, relation of $\Delta t \cdot \Delta E \cong \hbar$ holds between Δt and ΔE where ΔE is the energy width superimposed to make a wave packet and Δt is a time width in which the wave intensity (proportional to the square of the amplitude) is not zero.

5) **The last and fifth example**

The fifth example is a special case where the amplitude is constant during a certain period of time from $t = 0$ to $t = \tau_0$ and the amplitude is zero at other times

$$\begin{aligned} \Psi(t) &= \varphi_0 : \ \text{const} \quad 0 < t < \pi \\ \Psi(t) &= 0 : \quad t < 0 \quad \text{or} \quad t > \pi \end{aligned} \tag{4.26}$$

The Fourier component of Eq. (4.20) is obtained by using Formula (4.26) as

$$\begin{aligned} a(\nu) &= \int_{-\infty}^{\infty} \Psi(t)\exp(i2\pi\nu t)\mathrm{d}t = \int_0^{\tau_0} \varphi_0 \exp(i2\pi\nu t)\mathrm{d}t \\ &= \left[\frac{\varphi_0}{i2\pi\nu}\exp(i2\pi\nu t)\right]_0^{\tau_0} = \frac{\varphi_0}{i2\pi\nu}\left[\exp(i2\pi\nu\tau_0) - 1\right] \end{aligned} \tag{4.27}$$

Therefore,

$$\begin{aligned} |a(\nu)|^2 = a^*(\nu)a(\nu) &= \frac{\varphi_0^2}{4\pi^2\nu^2}(2 - \exp(i2\pi\nu\tau_0) - \exp(-i2\pi\nu\tau_0)) \\ &= (\varphi_0\tau_0)^2\left(\frac{\sin\pi\nu\tau_0}{\pi\nu\tau_0}\right)^2 \end{aligned} \tag{4.28}$$

The quantity $|a(\nu)|^2$ contains a type of function $(\sin\xi)/\xi$. The main local maximum of this function occurs at $\xi = 0$, where $(\sin\xi)/\xi = 1$. Furthermore, this becomes 0 when $\xi = \pm\pi, \pm 2\pi, \cdots, \pm m\pi$. In the middle, when the value of ξ satisfies the condition of $\tan\xi = \xi$, it has a secondary maximum. Most of the area surrounded by the function $(\sin\xi/\xi)^2$ and the horizontal axis is concentrated in the section between $\xi = -\pi$ and $+\pi$, and the area outside this section does not exceed 5% of the above total area. Therefore,

$$\Delta E = E_2 - E_1 = h\nu_2 - h\nu_1 = h\left(\frac{1}{\tau_0}\right) - \left(h \cdot \frac{-1}{\tau_0}\right) = \frac{2h}{\tau_0} \tag{4.29}$$

On the other hand, Δt, the duration time of the state, is given by

$$\Delta t = \tau_0 \tag{4.30}$$

Thus, $\Delta t \cdot \Delta E \cong \hbar$ is obtained from Eqs. (4.29) and (4.30). The fourth and fifth examples of Table 4.1 are for the case where the wave function Ψ is a non-periodic function of time, and thus, the spectrum is continuous, as expected.

From the examples shown in Table 4.1, the following conclusions can be drawn that serve as a basis for the formulation of new kinetics. *That is*, when there is no fluctuation in the energy, and thus a single definite value is given with $\Delta E = 0$, the duration or lifetime of the state of the system becomes infinite. When the lifetime is finite, the state of the system can be expressed as a superposition of states with different energies. This conclusion is extremely important. In the state of a system with a finite lifetime, we cannot describe the system as a state with a certain definite energy. Therefore, we must describe it as a superposition of different energy states. In short, the energy of the transition state which has a finite lifetime is uncertain, or indefinite; the lifetime, Δt, is determined by the width of uncertainty in energy, ΔE. Thus, it was again confirmed that the uncertainty relation $\Delta t \cdot \Delta E \cong \hbar$ should be the fundamental principle for constructing reaction kinetics.

4.2 Thermal Activation in Phase Transformations and Chemical Reactions

The problem of thermal activation has been one of the important issues in the kinetics of phase transformations and chemical reactions since the concept of activated molecules was proposed by Arrhenius. In this section, we will interpret this thermal activation problem based on the uncertainty relation. The basic principle is to explain the progress of the reaction caused by increasing the temperature by the fact that the energy fluctuation width increases with the increase in the temperature, and therefore, the lifetime of reactants or phases is shortened due to the uncertainty relation. Then, based on the explanation, we will discuss some aspects of thermal activation.

4.2.1 *Interpretation of Thermal Activation Based on Uncertainty Relation,* $\Delta t \cdot \Delta E \cong \hbar$

Suppose that the system is now in a stationary state. Then, as is clear from the discussion in the previous section, there is no energy fluctuation ($\Delta E = 0$) in the system, of which lifetime in that state is infinite. The state of such a system is represented by the solution φ_{n} of the time-independent wave equation. Assuming that the wave function φ_{n} is normalized, the mean energy $\langle E \rangle$ of the system is given by

$$\langle E \rangle = \int \varphi_n^* H \varphi_n \mathrm{d}q = E_n \int \varphi_n^* \varphi_n \mathrm{d}q = E_n \tag{4.31}$$

That is, when there is no fluctuation, the average value of energy is naturally equal to the eigenvalue.

Next, suppose the system is in a non-stationary state. In this case, the energy fluctuation is not zero ($\Delta E \neq 0$), and the lifetime of the system is finite. According to the conclusion of the previous section, the state of such a system can be expressed as a superposition of states with different energies, that is, the wave function representing the state of the system can be expressed by the following equation.

$$\Psi = \sum_n c_n \varphi_n \tag{4.32}$$

where φ_{n} is assumed to form an orthonormal system.

The mean energy is

$$\langle E \rangle = \int \Psi^* H \Psi \mathrm{d}q = \sum_{m,n} c_m^* c_n E_n \int \varphi_m^* \varphi_n \mathrm{d}q = \sum_n c_n^* c_n E_n = \sum_n w_n E_n \tag{4.33}$$

where $w_n \equiv C_n^* C_n$ is called the density matrix, and w_n is given by Eq. (1.32)

$$w_n = A \exp\left(-\frac{E_n}{kT}\right) \tag{4.34}$$

where A is a constant for normalization.

In order to further calculate the mean energy $\langle E \rangle$, the energy E_n must be given. By the way, we are considering the energy fluctuation caused by nonzero temperature. The energy is thought to fluctuate around minimum position of the potential. Then, let $x = a$ be the point where the potential becomes minimum. If we expand the potential V into a power series of $(x - a)$ near this point, we can approximate V as follows, since $\partial V / \partial x = 0$ at $x = a$.

$$V \cong \frac{K}{2}(x-a)^2 \tag{4.35}$$

where K is constant. This is exactly the potential of the harmonic oscillator. Thus, to a first approximation, we can regard the system as consisting of a huge number of harmonic oscillators. The energy of the harmonic oscillator is given by quantum mechanics as

$$E_n = \left(n+\frac{1}{2}\right)h\nu \quad (n = 0, 1, 2, \ldots) \tag{4.36}$$

Therefore,

$$\langle E \rangle = \sum_n w_n E_n = \frac{\sum_{n=0}^{\infty}\left(n+\frac{1}{2}\right)h\nu \exp\left[-\left(n+\frac{1}{2}\right)\left(\frac{h\nu}{kT}\right)\right]}{\sum_{n=0}^{\infty}\exp\left[-\left(n+\frac{1}{2}\right)\left(\frac{h\nu}{kT}\right)\right]} = \frac{h\nu}{2} + \frac{h\nu}{\exp\left(\frac{h\nu}{kT}\right)-1} \tag{4.37}$$

When $h\nu \ll kT$, we can rewrite Eq. (4.37) as

$$\langle E \rangle = kT \tag{4.38}$$

This means that the classical limit has been determined, and in fact the limit value kT is equal to the average energy of the harmonic oscillator in classical mechanics.

On the other hand, when $h\nu \gg kT$,

$$\langle E \rangle = \frac{h\nu}{2} \tag{4.39}$$

This corresponds to the fact that the oscillator has only zero-point energy at cryogenic temperatures.

The mean value of the square of the energy is

$$\begin{aligned}\left\langle E_n^2 \right\rangle &= \frac{\sum_{n=0}^{\infty}\left\{\left(n+\frac{1}{2}\right)h\nu\right\}^2 \exp\left[-\left(n+\frac{1}{2}\right)\left(\frac{h\nu}{kT}\right)\right]}{\sum_{n=0}^{\infty}\exp\left[-\left(n+\frac{1}{2}\right)\left(\frac{h\nu}{kT}\right)\right]} \\ &= (h\nu)^2\left[\exp\left(\frac{2h\nu}{kT}\right) + 6\exp\left(\frac{h\nu}{kT}\right) + 1\right]\left\{2\left[\exp\left(\frac{h\nu}{kT}\right) - 1\right]\right\}^{-2}\end{aligned} \tag{4.40}$$

By using Eqs. (4.37) and (4.40), the energy fluctuation Γ becomes

$$\begin{aligned}\langle \Gamma \rangle^2 &\equiv \left\langle (E_n - \langle E_n \rangle)^2 \right\rangle = \left\langle E_n^2 \right\rangle - \langle E_n \rangle^2 \\ &= (h\nu)^2 \cdot \exp\left(\frac{h\nu}{kT}\right) \cdot \left[\exp\left(\frac{h\nu}{kT}\right) - 1\right]^{-2}\end{aligned} \tag{4.41}$$

Rearranging the formula, it becomes

$$\langle \Gamma \rangle^2 = (h\nu)\frac{h\nu}{\left[\exp\left(\frac{h\nu}{kT}\right) - 1\right]} + \frac{(h\nu)^2}{\left[\exp\left(\frac{h\nu}{kT}\right) - 1\right]^2} = (h\nu)\left\langle E' \right\rangle + \left\langle E' \right\rangle^2 \quad (4.42)$$

$$\left\langle E' \right\rangle = (h\nu)\left[\exp\left(\frac{h\nu}{kT}\right) - 1\right]^{-1} \quad (4.43)$$

where $\left\langle E' \right\rangle$ is the average energy when the zero-point energy is ignored in Eq. (4.37). Equation (4.42) shows that the square of the energy fluctuation consists of two terms, the one for particle-like nature, $(((h\nu)\langle E \rangle'))$, and the other for wave-like nature, $(\left\langle E' \right\rangle^2)$. This means that the system under consideration has the particle–wave duality. Equation (4.41) is reduced as follows.

$$\text{In case of } h\nu \ll kT : (\Gamma)^2 = (kT)^2 = \left\langle E' \right\rangle^2 \quad \text{(wave nature)} \quad (4.44)$$

$$\text{In case of } h\nu \gg kT : (\Gamma)^2 = (h\nu)\left\langle E' \right\rangle = \quad \text{(particle nature)} \quad (4.45)$$

These are consistent with the classically well-known results on specific heat, which show that the wave nature is more pronounced at high temperatures and the particle nature is more pronounced at low temperatures, respectively. Furthermore, Eq. (4.44), for example, indicates that the average value of the energy is equal to the width of fluctuation in the energy. When the fluctuation is large, this equation means that it is inappropriate to represent the energy of the system only by its average value, and that the fluctuations should be taken into account. It should be noted that such situation as the equality of the average value of the energy and its fluctuations is intrinsic to the fact that the distribution of energetic states is given by Eq. (4.34).

Now, the energy fluctuation, $\Gamma$from Eq. (4.41) is given as

$$\Gamma = (h\nu)\left[\exp\left(\frac{h\nu/2}{kT}\right) - \exp\left(-\frac{h\nu/2}{kT}\right)\right]^{-1} \quad (4.46)$$

Therefore, the lifetime, τ is obtained from the uncertainty relation $\Gamma \cdot \tau \cong \hbar$:

$$\tau = \left(\frac{1}{2\pi\nu}\right)\left[\exp\left(\frac{h\nu/2}{kT}\right) - \exp\left(-\frac{h\nu/2}{kT}\right)\right] \quad (4.47)$$

Here, the conventional symbols of ΔE and Δt expressed for fluctuation width of energy and lifetime are replaced byΓ and τ, respectively, for future convenience. In many reactions, the situation $h\nu \gg kT$ holds. In this case, Eqs. (4.46) and (4.47) are reduced, respectively, as

$$\Gamma = (h\nu)\exp\left(-\frac{h\nu/2}{kT}\right) \tag{4.48}$$

Therefore,

$$\tau = \left(\frac{1}{2\pi\nu}\right)\exp\left(\frac{h\nu/2}{kT}\right) \tag{4.49}$$

At this point, we are ready to interpret thermal activation based on the uncertainty relation. The thermal activation in the phase transformation or the chemical reaction can be explained using Eqs. (4.46) and (4.47). That is, we know such phenomena that the rates of transformations from old phases to new phases or the rates of various chemical reactions from reactants to products increase as the temperature rises. These phenomena can be explained in such a way that the energy fluctuation of the old phases or the reactants increases according to Eq. (4.46), and therefore, the lifetime of the old phases or the reactants shortens according to Eq. (4.47). This is the explanation of the reason why the rates of various processes increase with rising temperature. Thus, we can now, in principle, interpret thermal activation in phase transformations or chemical reactions based on the uncertainty relation.

4.2.2 Lifetime and Reaction Rate

Next, let us turn to discussion about the relationship between the reaction rate and the lifetime τ in the uncertainty relation$\Gamma \cdot \tau \cong \hbar$. For this purpose, we will consider the law of decrease with time for atoms in the unstable or excited materials and the mean lifetime of atoms in the excited state.

Now, suppose that a system consists of a huge number of atoms, whose fluctuation width of energy is Γ and mean lifetime is τ, and the relationship between Γ and τ is given by $\Gamma{\cdot}\tau \cong \hbar$. We focus on any one atom in such unstable system. If an atom is still unstable at a certain moment t, in the following minute time dt, the atom may remain in the same state or transition to a more stable state with releasing extra energy. We shall consider it as a coincidental phenomenon. Suppose we can define the probability that an atom transitions to a more stable state in one second after time t, and let it be A.

Since it is an accidental phenomenon in a system consisting of a huge number of atoms, we cannot say which atom makes a transition between t and t +dt. However, we can predict how many atoms will undergo transition on average if we know the probability A. That is, if the number of atoms in an unstable state at time t is N, the number of transitions, – dN, between t and t + dt is given by,

$$-\mathrm{d}N = AN\mathrm{d}t \tag{4.50}$$

Integration of this equation gives the following equation.

$$N = N_0 e^{-At} \tag{4.51}$$

where N_0 is the number of unstable atoms at time $t = 0$.

Now let us calculate the mean lifetime in the unstable state as follows. The number of atoms transitioning between t and $t + \mathrm{d}t$ is equal to $AN\mathrm{d}t$. This is also the number of atoms that survive in an unstable state for t seconds. Therefore, the sum of their lifetimes is equal to $tAN\mathrm{d}t$, and the sum of the lifetimes of all atoms transitioning from $t = 0$ to $t = \infty$ is $\int_0^\infty tAN\mathrm{d}t$. The mean lifetime obtained from this is

$$\tau = \frac{1}{N_0}\int_0^\infty tAN\mathrm{d}t \tag{4.52}$$

Or, taking Eq. (4.51) into consideration, we obtain as follows.

$$\tau = A\int_0^\infty te^{-At}\mathrm{d}t \tag{4.53}$$

After performing partial integration and substituting the limit values,

$$\tau = \frac{1}{A} \tag{4.54}$$

By taking Eq. (4.54) into account, Eq. (4.51) can be rewritten as follows;

$$N = N_0 \exp[-(t/\tau)] \tag{4.55}$$

The time constant τ in Eq. (4.55) is sometimes called a relaxation time.

In any case, we have reached the following conclusions. If a system consists of a huge number of atoms, the fluctuation width of energy per one particle in the system is Γ, and its lifetime is $\tau = \hbar/\Gamma$, then the law of decrease in the number of unstable atoms obeys Eq. (4.55) with the lifetime τ as a time constant.

4.2.3 Derivation of Arrhenius Equation

According to Arrhenius, the temperature dependence of the reaction rate can be empirically expressed by the following specific reaction rate.

$$\kappa = A\exp\left[-\left(\frac{E}{kT}\right)\right] \tag{4.56}$$

A is called a frequency factor and has a dimension of frequency. On the other hand, E is called the activation energy that is independent of temperature unless the temperature range of the experiment is too wide.

Eyring's absolute reaction kinetics is, as already described in Chap. 2, trying to theoretically derive E and A in Equation (4.56). From our standpoint, the equation corresponding to Equation (4.56) is derived as follows. The specific rate constant κ is defined as follows.

$$\text{(reaction rate)} = \kappa \text{ (concentration of reactant)} \tag{4.57}$$

According to this definition, we can easily determine the specific rate constant κ. Differentiating Eq. (4.55) with respect to time, the reaction rate in this case is as follows.

$$-\frac{\mathrm{d}N}{\mathrm{d}t} = \frac{1}{\tau} N_0 \exp[-(t/\tau)] = \frac{1}{\tau} N \tag{4.58}$$

Comparing Eqs. (4.57) and (4.58) gives $\kappa = 1/\tau$, and τ, which is a function of temperature, is given by Eq. (4.47). That is,

$$\kappa = \frac{1}{\tau} = (2\pi\nu)\left[\exp\left(\frac{h\nu/2}{kT}\right) - \exp\left(-\frac{h\nu/2}{kT}\right)\right]^{-1} \tag{4.59}$$

In many cases, $h\nu \gg kT$ is satisfied, and in this case, Equation (4.59) is

$$\kappa = (2\pi\nu) \exp\left(-\frac{h\nu/2}{kT}\right) \tag{4.60}$$

From the correspondence between Eqs. (4.56) and (4.60), A and E are given as follows,

$$A = 2\pi\nu,\ E = h\nu/2 \tag{4.61}$$

A has a dimension of the frequency as expected, and E is energy that is explicitly independent of temperature. The term $h\nu$ is originally derived from Eq. (4.37) and is the interval between quantized energy levels in case of the harmonic oscillator approximation.

4.2.4 Critique of Concept of Thermal Activation and Arrhenius Equation

Thermodynamical Interpretation for Arrhenius Equation

Various interpretations have been made for the **Arrhenius Eq.** (4.56). Among them, the thermodynamic interpretation is as follows. Now let us consider the reaction

$$A + B \rightarrow C \tag{4.62}$$

where A and B react to produce C.

When this reaction is in equilibrium, the rightward and leftward reaction rates are equal. Therefore, in the case where the reaction rate is proportional to the simple concentration products,

$$\kappa_1 C_A C_B = \kappa_2 C_C,\ K_C = \frac{\kappa_1}{\kappa_2} = \frac{C_C}{C_A C_B} \tag{4.63}$$

Here, κ_1 and κ_2 are the reaction rate constants, K_C is the equilibrium constant, and C_A, C_B, and C_C, represent the concentrations of A, B, and C, respectively. On the other hand, according to thermodynamics, the equation of isochoric equilibrium of reaction (4.62) is

$$\frac{d \ln K_C}{dT} = \frac{\Delta E}{kT^2} \tag{4.64}$$

where ΔE is heat of reaction at constant volume.

Now, suppose that from the comparison between Eqs. (4.63) and (4.64) under the relation of $\Delta E = E_1 - E_2$ can lead to the formulation:

$$\frac{d \ln K_C}{dT} = \frac{d \ln \kappa_1}{dT} - \frac{d \ln \kappa_2}{dT} = \frac{E_1}{kT^2} - \frac{E_2}{kT^2} \tag{4.65}$$

In other words, let us make the assumption as

$$\frac{d \ln \kappa_1}{dT} = \frac{E_1}{kT^2} \quad \text{and} \quad \frac{d \ln \kappa_2}{dT} = \frac{E_2}{kT^2} \tag{4.66}$$

Let us regard E_1 and E_2 as constants and perform integration of Eq. (4.66), and we can obtain the following expressions with similar form of Eq. (4.56).

$$\kappa_1 = A_1 \exp\left[-\left(\frac{E_1}{kT}\right)\right] \text{ and } \kappa_2 = A_2 \exp\left[-\left(\frac{E_2}{kT}\right)\right] \tag{4.67}$$

Conventional Interpretation for Activation Energy

1) the following thermodynamical interpretation has been given to the activation energy. When A and B react to produce C, the reaction process can be represented as shown in Fig. 4.2, assuming that the product C is formed after passing through a high energy state on the way. E_1 is the difference between the energy that the molecule must possess at least for a rightward reaction to occur and the energy that the molecule has at absolute zero degree. E_2 is a similar energy difference for the reverse reaction. Usually, E_1 and E_2 are called activation energies for

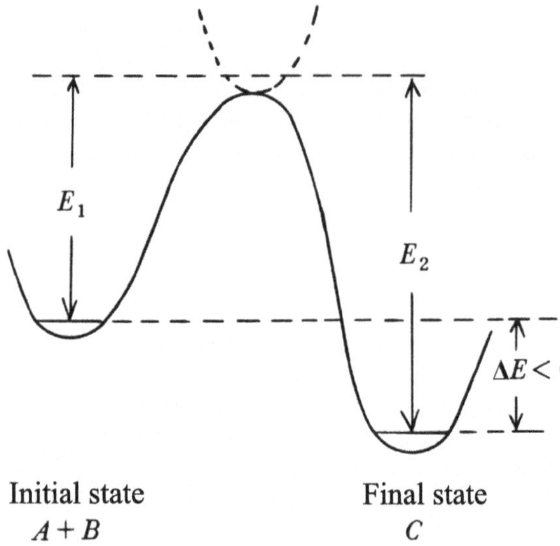

Fig. 4.2 Activation energies: E_1 for rightward reaction and E_2 for reverse reaction

rightward and leftward reactions, respectively. The similar concept is applied to other quantities of thermodynamics, such as enthalpy, entropy, and free energy, which lead to the concepts such as activation enthalpy, activation entropy, and activation free energy.

2) Tolman interpreted activation energy from the standpoint of statistical mechanics. According to his idea, the activation energy E_a is the difference between the average energy $\langle E^* \rangle$ of all molecules involved in the reaction and the average energy $\langle E \rangle$ of all molecules including those not involved in the reaction. That is,

$$E_a = \left\langle E^* \right\rangle - \langle E \rangle \tag{4.68}$$

This means that molecules involved in the reaction at a given temperature must have an energy greater than the average value by an amount of activation energy in advance.

3) another interpretation was given by Eyring. As described in Chap. 2, the difference between the potential energy of the activated complex and that of the initial state is considered as activation energy. The difference between the thermodynamic interpretation of the activation energy shown in Fig. 4.2 and Eyring's quantum mechanical one lies in the following point. The activation energy in the thermodynamic interpretation is, of course, the thermodynamic energy, while Eyring's interpretation of the activation energy is the eigenvalue E of the time-independent wave Eq. (3.3). Therefore, it has a different character from thermodynamic energy (see Chap. 1; Sect. 1.8.4 and Sect. 1.8.12).

So far, we have reviewed the conventional interpretation of activation energy in the process of thermal activation. Next, let us consider the problems about thermal activation and the Arrhenius equation.

Problems of Conventional Interpretations

1) **Definiteness of energy in transition state**

The first issue that should be addressed is the definiteness of energy in the transition state of the reaction caused by thermal activation. In every interpretation about activation energy described above, it is assumed that the system has a definite value in energy even in the transition state. However, as already discussed in Sect. 3.1.3, the energy is uncertain with fluctuation in any states having finite lifetimes, not only in the transition state but also in the states of systems before reaction. The uncertainty relation $\Gamma \cdot \tau \cong \hbar$ is established between the uncertainty width in energy Γ and the lifetime τ. Therefore, energy should not be treated as a definite quantity with certainty as in the conventional interpretation. Since the duration of the transition state is short, the width of the uncertainty is particularly large, and it has already been pointed out that the magnitude is about the same as so-called activation energy or reaction energy, and it is clear as shown in Table 3.1. The single-valuedness of energy in the transition state is assumed in such a way of saying that the transition state has an intrinsic internal energy depending only on the state of a system, but this assumption is unacceptable.

2) **Uncertainty of energy in transition state**

Since the degree of uncertainty in energy of the transition state has nearly the same order of magnitude as so-called activation energy or the reaction energy itself, it is not meaningful to claim that the transition state has a higher energy than the initial or final state, as shown in Fig. 4.2. What should be argued about the transition state is that the width of uncertainty in energy is extremely large and its magnitude is even comparable to the so-called activation energy or the reaction energy. The uncertainty of energy in the transition state is generated by the wave–particle duality of matter. In the electron-atom reaction in which wave–particle duality is important, as described in Sect. 3.1.3, consideration of only the height of energy peak for overcoming the energy barrier is not sufficient. Resonance transmission with a resonance energy should also be considered. Therefore, secondly, it is not sufficient to explain the activation energy in such a way of classical mechanics as whether the energy exceeds the barrier or not, but the quantum effect due to the duality of matter should be taken into account for interpreting the activation energy.

3) **Nature of activation energy**

Thirdly, let us consider the nature of the activation energy E in the Arrhenius Eq. (4.56). According to the conventional interpretation, this activation energy is defined as the difference between the energy in the activated state and the energy in the state before the reaction, but this definition does not hold since the energy in the activated state is not definite as mentioned above. In addition, there is another problem in terms of the activation energy E as the property of physical quantities.

(1) **First,** as already mentioned, the activation energy in the thermodynamical interpretation is, of course, the thermodynamic quantity. Therefore, it must have general characters as thermodynamic quantities described in Sect. 1.8.12. The important natures related to the concept of thermal activation are as follows.

  (a) Thermodynamic quantity is related to a macroscopic one.
  (b) Thermodynamic quantity is a state quantity defined for the equilibrium state, and therefore, energy E is a function of temperature and pressure.

  From the point described in (i), thermodynamic energy does not seem as a suitable physical quantity to express activation energy, because we are dealing with such a microscopic phenomenon as thermal activation of individual atoms or molecules. Also from the characteristics stated in (ii), it is impossible to regard the activated state as an equilibrium one. The energy in thermodynamics is a function of temperature and pressure, whereas the Arrhenius-type activation energy is independent of temperature so long as the temperature range is not very wide.

(2) **Next,** in Tolman's interpretation, activation energy is originally interpreted as molecular energy, so activation energy is considered to be a microscopic quantity. However, as shown in Eq. (4.68), the activation energy is the difference between the average value of the energy of all the molecules involved in the reaction and the average energy of all the molecules including the molecules not involved in the reaction. Since the average energy of a molecule is significantly dependent on temperature, the activation energy will also be dependent on temperature.
(3) **Finally,** in Eyring's interpretation, the activation energy is the difference in potential energy, and the potential energy is obtained as the eigenvalue of the wave equation in his theory; that is, the activation energy is a microscopic quantity. Since Eyring's activation energy is the difference between the potential energies of activated complex and initial state, it is not an explicit function of temperature, and his idea meets the requirement of Arrhenius-type activation energy. However, as already pointed out as the first problem, his idea involves a contradiction in the more fundamental level, that is, in his treatment that the energy has a definite value without uncertainty even in the transition state.

4) **Assumption of equilibrium**

The fourth problem is the assumption of equilibrium in defining the activation energy. In the thermodynamic interpretation of activation energy, the definition of energy itself is based on the assumption of equilibrium. Also, for the thermodynamical formulation in Eyring's absolute reaction kinetics, as mentioned in Chap. 2, a chemical equilibrium is assumed between the reactants and the activated complex. It is quite unacceptable to regard the transition state as an equilibrium state, or to assume even approximately the equilibrium in the progressing reaction, since those assumptions introduce a crucial contradiction into the logical structure of constructing physical theory.

Explanation in Our Theory

However, **first of all**, we do not assume in our treatment that the energy is definite without uncertainty in the transition state. Conversely, we consider it as uncertain and develop the theory based on its quantitative expression, the uncertainty relation $\Gamma \cdot \tau \cong \hbar$.

Secondly, quantum effects when crossing the energy barrier are well taken into consideration by quantum mechanical treatment.

Thirdly, the physical quantity representing the activation energy, $h\nu/2$ (Eq. (4.61)), corresponding to the activation energy is a microscopic quantity, and $h\nu$ is the interval between the quantized energy levels and independent of temperature.

Fourthly, regarding the assumption of equilibrium, $h\nu$ itself is the interval between the energy levels of the old phase or the reactant before the reaction, and therefore, our theoretical treatment has no relation to the transition state or activated complexes.

What we are concerned with is the energy fluctuations and lifetimes of the old phases or reactants. There is no need to assume equilibrium for the processing reaction itself; the assumption of equilibrium for the transition state or the assumption of chemical equilibrium between the reactant and the activated complex is not necessary at all.

References

1. D. Bohm, *Quantum Theory* (Prentice-Hall Inc., 1951)
2. A. Messiah, *Mécanique Quantique, (Dunod, 1959), (English version, Quantum Mechanics*, vol. 1 (North-Holland Publishing, Amsterdam/Oxford, 1975)

Further Reading

W.C. Price, S.S. Chissik, *The Uncertainty Principle and Foundations of Quantum Mechanics* (John Wiley and Sons, A fifty years' survey, 1977)

Part III
Application and Characteristics of Our Reaction Kinetics

Chapter 5
Application of Our Reaction Kinetics to Simple Systems

Abstract In this chapter, the physical formulation and the mathematical formulation developed in previous chapters are first applied to solve various elementary problems such as diffusion, melting, and boiling. More complex phenomena such as precipitation, graphitization, and chemical reactions are then addressed. An interpretation of thermally activation process based on the uncertainty principle provides a satisfactory explanation of diffusion phenomena and melting and boiling processes of metals. Next, the physical meaning of the phase transformation equation formulated by Johnson-Mehl will be clarified. However, the Johnson-Mehl equation is applicable only when the number of particles is constant or the nucleation rate of particle is constant. We present a newly generalized rate equation that is applicable when the nucleation rate of new-phase particles increases or decreases exponentially. Furthermore, the graphitization of cementite by impact deformation is an example of reactions by mechanical actions.

Keywords Diffusion · Melting point · Boiling point · Precipitation · Graphitization · Impact deformation · Chemical reaction · Johnson–Mehl equation · Thermal activation

In this chapter, the physical ideas developed in Chap. 3 and the mathematical ideas developed in Chap. 4 are applied to solving various elementary problems. From easiest case to more complicated ones will be explained in order. We apply our reaction kinetics firstly to the phenomena of diffusion and melting/boiling, which are seemed to be the most basic and elemental phenomena for analyzing phase transformations or chemical reactions. After that, let us apply our theory to more complex phenomena, such as precipitation, graphitization, and chemical reactions.

S. Yamamoto, *Reaction Kinetics Based on Time-Energy Uncertainty Principle*,
https://doi.org/10.1007/978-981-19-9673-3_5

5.1 Diffusion

Let us start by considering the problem of one-dimensional random walk. The one-dimensional coordinate is divided at equal interval a, and numbers are assigned as shown in Fig. 5.1.

It is assumed that a point s that advances on this coordinate is initially located at the coordinate 0. Now, when a fair coin with the same probability of appearing heads or tails is flipped, if the head comes out, s is advanced to right by one division, and if the tail comes out, it is advanced to left by one division. Each trial is an independent event because the probability whether heads or tails occur on each toss does not depend on the result of the previous trial. Since s has two ways to go left or right for one trial, there are a total of 2^N routes that s can take in successive N trials. Therefore, the probability of taking one route of them is $1/2^N$. In order to find s at point m after N trials, the difference between the number of heads and tails must be m, so it is limited to the case where the heads appear $(N+m)/2$ times and tails appear $(N-m)/2$ times. The total number of combinations that achieve this case is $N!/\{[(N+m)/2]![(N-m)/2]!\}$. Being even or odd for m is decided whether N is even or odd. The probability $P(m, N)$ that s reaches the point m after N trials in this way is expressed as:

$$P(m, N) = \frac{N!}{\left[\frac{1}{2}(N+m)\right]!\left[\frac{1}{2}(N-m)\right]!}\left(\frac{1}{2}\right)^N \tag{5.1}$$

Since the head and tail of the coin have equal probability, the average value of m can be expected to be 0, when N is increased. In fact, $P(m, N)$ rapidly approaches 0 where m is large, so it is sufficient to consider the region of $m \ll N$ as a practically important case.

Here, by taking logarithm of both sides of Eq. (5.1) and using the Stirling's formula for very large n:

$$\ln n! = \left(n + \frac{1}{2}\right)\ln n - n + \left(\frac{1}{2}\right)\ln 2\pi + \mathrm{O}\left(n^{-1}\right)(n \to \infty) \tag{5.2}$$

we will have

$$\ln P(m, N) \cong \left(n + \frac{1}{2}\right)\ln N - \frac{1}{2}(N + m + 1)\ln\left\{\frac{N}{2}\left(1 + \frac{m}{N}\right)\right\}$$

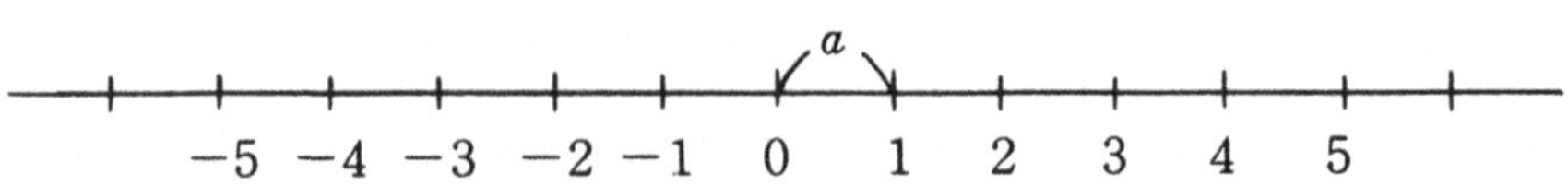

Fig. 5.1 One-dimensional model of random walk

$$-\frac{1}{2}(N-m+1)\ln\left\{\frac{N}{2}\left(1-\frac{m}{N}\right)\right\}-\frac{1}{2}\ln 2\pi - N\ln 2 \quad (5.3)$$

When *m/N* is small enough, using the equation:

$$\ln(1+x) = x - \frac{x^2}{2} + O\left(x^3\right) \quad (5.4)$$

Equation (5.3) is rewritten as

$$\ln P(m, N) \cong -\frac{1}{2}\ln N + \ln 2 - \frac{1}{2}\ln 2\pi - \frac{m^2}{2N} \quad (5.5)$$

By converting Eq. (5.5) to an exponent-type function, we obtain

$$P(m, N) = \left(\frac{2}{\pi N}\right)^{1/2} \exp\left(-\frac{m^2}{2N}\right) \ (m \ll N) \quad (5.6)$$

Let $r = ma$ be the distance that s has moved from the origin 0. If we take dr to be sufficiently larger than the interval a, the probability of finding s in $r \sim r + \mathrm{d}r$ becomes

$$P(r, N)\mathrm{d}r = \frac{P(m, N)\mathrm{d}r}{2a} \quad (5.7)$$

Here, the reason why the denominator on the right side of Eq. (5.7) is 2 is that m can take only either even or odd depending on N. Using this equation, Eq. (5.6) is expressed as

$$P(r, N) = \frac{1}{\sqrt{2\pi N a^2}}\exp\left(-\frac{r^2}{2Na^2}\right) \quad (5.8)$$

In the case of n trials per unit time, we can write $N = nt$, so the probability of finding s in the region $r \sim r + \mathrm{d}r$ at time t can be expressed by

$$P(r, t)\mathrm{d}r = \frac{1}{2\sqrt{\pi D t}}\exp\left(-\frac{r^2}{4Dt}\right)\mathrm{d}r \quad (5.9)$$

where D is given by

$$D = \frac{na^2}{2} \quad (5.10)$$

This is an example of the central limit theorem saying that "when independent random variables are added, the normalized sum tends to approach a normal distribution". As long as the displacement is represented by the sum of independent small displacements, such as Brownian motion and diffusion of the particles that undergo displacement due to the collision of particles in a heat bath, the probability distribution of the displacement after a sufficiently long time t can be expressed by Eqs. (5.9) and (5.10).

Now, by using the results of the interpretation based on the uncertainty relation for the thermal activation described earlier, the temperature dependence of the diffusion coefficient D given by Eq. (5.10) can be expressed explicitly. For example, let us consider diffusion occurring at temperature T. The time that the atom of interest stays at a certain position is given by τ in Eq. (4.47). Therefore, n in Eq. (5.10), i.e., the number of jumps per unit time is given by the reciprocal of the lifetime τ of the atom remaining at a certain position. Therefore, Eq. (5.10) is expressed as

$$D = \frac{\pi \nu a^2}{\left\{\exp\left(\frac{h\nu/2}{kT}\right) - \exp\left(-\frac{h\nu/2}{kT}\right)\right\}} \tag{5.11}$$

As is usually the case, $h\nu \gg kT$. In this case, Eq. (5.11) becomes

$$D = \pi \nu a^2 \exp\left(-\frac{h\nu/2}{kT}\right) \tag{5.12}$$

or

$$D = D_0 \exp\left(-\frac{E}{kT}\right) \tag{5.13}$$

where

$$E = \frac{h\nu}{2} \quad \text{and} \quad D_0 = \pi \nu a^2 = 2\pi a^2 \frac{E}{h} \tag{5.14}$$

From Eq. (5.14), the following two conclusions can be drawn. First, the activation energy of diffusion is equivalent to $h\nu/2$, and therefore, if the activation energy of diffusion is known, the coefficient D_0 can be evaluated. Conversely, if the coefficient D_0 is known, the activation energy can be evaluated. Second, the coefficient D_0 depends on the activation energy and becomes larger with the increase in activation energy.

Next, we will compare the conclusion drawn from Eq. (5.14) with experimental data and examine its validity. Table 5.1 shows the experimental data of diffusion and the calculated value of D_0 by Eq. (5.14). In calculating D_0, the experimental values corresponding the activation energy were used [1, 2]. The value a is the interatomic distance. The diffusion data in Table 5.1 shows that the experimental values vary considerably depending on the experimental method and other conditions. Despite the data having such variations, it can be seen that Eq. (5.14) including the universal constant called Planck's constant gives a fairly good agreement within the range of the data variations. Moreover, looking at the changes in the activation energy and diffusion coefficient D_0 for the self-diffusion in α-Fe, γ-Fe, and δ-Fe, and also for the diffusion of the substitutional and interstitial impurities, the diffusion coefficient D_0 increases when the activation energy becomes larger, which is consistent with the change predicted by Eq. (5.14). Thus, the interpretation of thermal activation by the uncertainty principle seems in principle to be correct in the diffusion phenomenon. It should be noted that this conclusion also applies to diffusion data for elements other than iron, such as Cu, Ag, and Au.

5.2 Melting and Boiling of Metals

As a second example of applying the interpretation of thermal activation by the uncertainty principle, we consider the melting and boiling of metals. How can these phenomena be interpreted from our standpoint? To proceed with the analysis for clarifying those phenomena, we will use the following equations of energy fluctuation Γ and lifetime τ during the thermal activation process obtained in the previous Chap. 4.

$$\Gamma = \frac{h\nu}{\left\{\exp\left(\frac{h\nu/2}{kT}\right) - \exp\left(-\frac{h\nu/2}{kT}\right)\right\}} \tag{5.15}$$

$$\tau = \frac{1}{2\pi\nu}\left\{\exp\left(\frac{h\nu/2}{kT}\right) - \exp\left(-\frac{h\nu/2}{kT}\right)\right\} \tag{5.16}$$

In these equations, if the latent heat of melting (boiling) is given as $h\nu$ and the melting point (boiling point) is given as temperature T, Γ and τ should be the energy fluctuation and the lifetime in melting (boiling), respectively. Table 5.2 and Fig. 5.2 show the analysis results of the melting of metals, and Table 5.3 and Fig. 5.3 show those for the boiling of metals, respectively.

The results of the above analysis for the melting and boiling of metals are of great interest. First of all, despite the fact that melting (boiling) of metals occurs over a wide range of melting points (boiling points) and latent heat in accordance with wide variations in the types of metals, it can be seen that they have almost constant lifetimes of 3.3×10^{-14} s (1.5×10^{-13} s). Each of these constant lifetimes can be understood as follows. In melting, the electron-mediated binding state changes from that of solid to that of liquid. Therefore, the constant lifetime of 3.3×10^{-14} s seems to correspond to the motion period of electrons. In other words, the energy fluctuation increases as the temperature rises, and the lifetime of the metal atoms that can remain in the state of solid becomes shorter. However, at the moment when the mean lifetime becomes equal to the characteristic period of electrons, a kind of resonance occurs, that is, the process of melting seems to proceed like an avalanche as a cooperative phenomenon.

Table 5.1 Comparison of theoretical and experimental values of diffusion coefficient [1, 2]

(1) Self-diffusion of α-Fe (interatomic distance, $a = 2.482$ Å)

Experimental data				Calculated value
Year	Temp. range of experiment (°C)	Activation energy kcal/mol (eV)	Diffusion coefficient D_0 (cm^2/s)	Diffusion coefficient D_0 (cm^2/s)
1948		52.00 (2.26)	1.4×10^{-1}	2.12
1948		77.2 (3.36)	3.4×10^{4}	3.14
1949		62.00 (2.70)	14.0	2.53
1950	720–887	73.2 (3.18)	2.3×10^{3}	2.98
1951	810–910	59.7 (2.60)	5.8	2.43
1952		48.0 (2.09)	0.10	1.96
1952		59.8 (2.60)	6.2	2.43
1955	650–850	67.1 (2.92)	5.3×10^{2}	2.73
1955	700–850	47.1 (2.05)	1.9×10^{-2}	1.92
1957	705–900	56.5 (2.46)	3.2	2.30
1958	775–892	64.1 (2.79)	18	2.61
1958		47.0 (2.04)	1.15×10^{-2}	1.92
1960	796–895	67.2 (2.92)	1.18×10^{2}	2.73
1960	775–890	64.1 (2.79)	18	2.61
1961	700–750 ferromagnetism	60.0 (2.61)	2	2.44
1966	683–726 ferromagnetism	60.7 (2.64)	27.5	2.47
1961	809–905 paramagnetism	57.2 (2.49)	1.9	2.33
1963	863–899 paramagnetism	57.3 (2.49)	2	2.33
1966	767.5–884 paramagnetism	57.5 (2.50)	2.01	2.34
1965	501.5–682 Grain boundary diffusion	41.5 (1.80)	1.12×10^{-6}	1.69

(continued)

Table 5.1 (continued)

(2) Self-diffusion of γ-Fe (interatomic distance, $a = 2.520$ Å)

Experimental data				Calculated value
Year	Temp. range of experiment (°C)	Activation energy kcal/mol (eV)	Diffusion coefficient D_0 (cm^2/s)	Diffusion coefficient D_0 (cm^2/s)
1950	970–1357	74.2 (3.23)	58.0	3.12
1951		67.9 (2.95)	0.58	2.85
1963		68.0 (2.96)	0.76	2.85
1955		67.9 (2.95)	1.3	2.85
1956	1000–1300	67.0 (2.91)	0.44	2.81
1957		56.5 (2.46)	3.2	2.37
1958		63.5 (2.76)	0.11	2.66
1961	1064–1349	64.5 (2.80)	0.18	2.71
1861	918–1259	60.0 (2.61)	1.85×10^{-2}	2.52
1862		67.7 (2.94)	2.46	2.84
1962		60.0 (2.61)	1.8×10^{-2}	2.52
1963	1156–1349	64.0 (2.78)	0.22	2.69
1965	1031–1320	67.8 (2.95)	1.05	2.84
1965	949–1159 Grain boundary diffusion	38.0 (1.65)	0.77×10^{-7}	1.59

(3) Self-diffusion of δ-Fe (interatomic distance, $a = 2.534$ Å)

Experimental data				Calculated value
Year	Temp. range of experiment (°C)	Activation energy kcal/mol (eV)	Diffusion coefficient D_0 (cm^2/s)	Diffusion coefficient D_0 (cm^2/s)
1961	1404–1518	42.4 (1.84)	1.9×10^{-2}	1.80
1962	1413–1507	57.0 (2.48)	1.9	2.42
1963	1407–1515	61.7 (2.68)	6.3	2.62
1964	1405–1515	60.8 (2.64)	8.3	2.58
1966	1428–1492	57.5 (2.50)	2.01	2.44

(continued)

Table 5.1 (continued)

(4) Diffusion of impurity in α-Fe (interatomic distance, $a = 2.482$ Å)

Experimental data					Calculated value
Year	Diffusion species	Temp. range of experiment (°C)	Activation energy kcal/mol (eV)	Diffusion coefficient D_0 (cm^2/s)	Diffusion coefficient D_0 (cm^2/s)
1968	Al	1000–1200	39.2 (1.70)	1.4×10^{-2}	1.60
1963	Au	700–900	64.2 (2.79)	31	2.61
1966	Co	683–726	62.20 (2.70)	7.19	2.53
1966	Cu	650–750	58.38 (2.54)	0.47	2.38
1955	Cr	750–850	82.0 (3.57)	3×10^4	3.34
1966	Mo	750–850	49 (2.13)	0.3	1.99
1961	Ni	810–900	56.0 (2.43)	1.3	2.28
1963	P	860–900	55.0 (2.39)	2.9	2.24
1959	Ti	1075–1225	59.2 (2.57)	3.15	2.41
1955	W	700–785	70.0 (3.04)	3.8×10^2	2.85

(5) Diffusion of impurity in γ-Fe (interatomic distance, $a = 2.520$ Å)

Experimental data					Calculated value
Year	Diffusion species	Temp. range of experiment (°C)	Activation energy kcal/mol (eV)	Diffusion coefficient D_0 (cm^2/s)	Diffusion coefficient D_0 (cm^2/s)
1962	Al		56.0 (2.43)	30	2.35
1961	Co	1140–1340	72.9 (3.17)	1.25	3.06
1955	Cr	1050–1250	97.0 (4.22)	1.8×10^4	4.07
1966	Cu	800–1200	61.0 (2.65)	3.0	2.56
1945	Mo		57.7 (2.51)	34.0	2.42
1959	Ni	1152–1400	77.6 (3.37)	6.92	3.26
1959	Ti	1075–1225	60.0 (2.61)	0.15	2.52
1955	W		75.0 (3.26)	13	3.15

(continued)

Table 5.1 (continued)

(6) Diffusion of impurity in δ-Fe (interatomic distance, $a = 2.534$ Å)					
Experimental data					Calculated value
Year	Diffusion species	Temp. range of experiment (°C)	Activation energy kcal/mol (eV)	Diffusion coefficient D_0 (cm^2/s)	Diffusion coefficient D_0 (cm^2/s)
1963	Co	1396–1502	61.2 (2.66)	5.5	2.60
1966	Co	1429–1521	61.4 (2.67)	6.38	2.60
1963	P	1370–1460	55.0 (2.39)	2.9	2.33
(7) Diffsion of carbon in α-Fe (interatomic distance, $a = 1.433$ Å)					
Experimental data					Calculated value
Year	Method	Temp. range of experiment (°C)	Activation energy kcal/mol (eV)	Diffusion coefficient D_0 (cm^2/s)	Diffusion coefficient D_0 (cm^2/s)
1950	Internal friction	− 35–200	20.1 (0.87)	2×10^{-2}	0.27
1949	Internal friction	0–200	19.8 (0.86)	8×10^{-3}	0.27
1954	Internal friction	24–74	20.4 (0.89)	3.16×10^{-2}	0.28
1955	C-14	350–850	18.5 (0.80)	4.2×10^{-3}	0.25
1949	Diffusion coupling	514–786	18.1 (0.79)	7.9×10^{-3}	0.25
1949	Concentration gradient	700–900	16.8 (0.73)	3.8×10^{-3}	0.23
1964	Internal friction		20.4 (0.89)	2.6×10^{-2}	0.28
(8) Diffusion of carbon in γ-Fe (interatomic distance, $a = 1.781$ Å)					
Experimental data					Calculated value
Year	Method	Temp. range of experiment (°C)	Activation energy kcal/mol (eV)	Diffusion coefficient D_0 (cm^2/s)	Diffusion coefficient D_0 (cm^2/s)
1955	C-14	900–1060	27.0 (1.17)	1×10^{-2}	0.56
1961	C-14		32.0 (1.39)	1.5×10^{-1}	0.67
1955	C-14	900–1060	32.4 (1.41)	1×10^{-1}	0.68
	Decarburization	850–1100	37.46 (1.63)	6.68×10^{-1}	0.78

(continued)

Table 5.1 (continued)

(9) Diffusion of boron in Fe

Experimental data					Calculated value
Year	Phase	Temp. range of experiment (°C)	Activation Energy kcal/mol (eV)	Diffusion Coefficient D_0 (cm²/s)	Diffusion Coefficient D_0 (cm²/s)
1955	α		15 (0.65)	1×10^{-5}	0.61
1953	γ	950–1300	21.6 (0.94)	1×10^{-3}	0.88

(10) Diffusion of nitrogen in Fe

Experimental data						Calculated value
Year	Phase	Method	Temp. range of experiment (°C)	Activation energy kcal/mol (eV)	Diffusion coefficient D_0 (cm²/s)	Diffusion coefficient D_0 (cm²/s)
1949	α	Internal friction	0–200	17.7 (0.77)	1.4×10^{-2}	0.72
1950	α	Internal friction	– 20–50	18.2 (0.79)	3×10^{-3}	0.74
1954	α			18.9 (0.82)	6.6×10^{-3}	0.77
1956	α		400–600	15.6 (0.68)	1.2×10^{-3}	0.63
1964	α	Dissolution rate	750–873	18.9 (0.82)	7.8×10^{-3}	0.77
1964	α	Dissolution rate	1410–1470	18.9 (0.82)	7.8×10^{-3}	0.77
1926	γ	Concentration gradient	1100	34.0 (1.48)	1.1×10^{-1}	1.43
1964	γ	Dissolution rate	1000–1202	40.26 (1.75)	9.1×10^{-1}	1.69

(11) Diffusion of oxygen in Fe

Experimental data						Calculated value
Year	Phase	Method	Temp. range of experiment (°C)	Activation energy kcal/mol (eV)	Diffusion coefficient D_0 (cm²/s)	Diffusion coefficient D_0 (cm²/s)
1966	α	Transmission	700–900	10.1 (0.44)	0.24	0.41
1967	δ	Transmission	1450–1510	10.1 (0.44)	0.24	0.43
1967	γ	Transmission	1033–1352	17.5 (0.76)	2.14	0.73

(continued)

Table 5.1 (continued)

(12) Diffusion of hydrogen α-Fe (interatomic distance, $a = 2.482$ Å)

Experimental data					Calculated value
Year	Method	Temp. range of experiment (°C)	Activation energy kcal/mol (eV)	Diffusion coefficient D_0 (cm^2/s)	Diffusion coefficient D_0 (cm^2/s)
1947	Transmission	400–900	2.28 (0.10)	7.6×10^{-4}	0.09
1950	Transmission	400–900	2.90 (0.13)	2.2×10^{-3}	0.12
1956	Emission	150–650	3.05 (0.13)	8.8×10^{-4}	0.12
1958	Emission	200–774	2.7 (0.12)	9.3×10^{-4}	0.11
1960	Emission	200–780	3.2 (0.14)	1.4×10^{-3}	0.13
1960	Absorption	430–725	2.2 (0.10)	2.2×10^{-4}	0.09
1957	Transmission (unsteady)	30–90	3.3 (0.14)	2.1×10^{-3}	0.13
1957	Absorption (unsteady)	26–90	3.4 (0.15)	5×10^{-3}	0.14
1957	Emission (unsteady)	26–90	6.32 (0.27)	1.9×10^{-1}	0.26
1960	Emission	25–200	7.82 (0.34)	1.2×10^{-1}	0.32

Annotation: The experimental data of diffusion in Table 5.1 is taken from Appendix in Japanese edition of "The Structure of Alloys of Iron: An Elementary Introduction" translated by K. Hirano [1]

Table 5.2 Analysis of melting of metals; *Databook of Metals* [3] and *Metallurgical Thermochemistry* [4]

Element	Melting point, T_m, K	Thermal energy, kT_m, eV	Latent heat, $h\nu$, kcal/mol (eV)	Fluctuation of energy, Γ, eV	Lifetime, τ, $\times 10^{-15}$ s
Ag	1233.8	0.106	2.69 (0.12)	0.101	6.25
Al	932	0.08	2.5 (0.11)	0.074	8.85
Au	1336	0.115	3.05 (0.13)	0.109	6.03
Ba	983	0.085	1.83 (0.08)	0.082	8.06
Be	1556	0.134	2.8 (0.12)	0.13	5.07
Bi	544.3	0.047	2.60 (0.11)	0.038	17.5
C	(4973)	(0.429)	(33) (1.43)	0.28	2.35
Ca	1123	0.097	2.07 (0.09)	0.0934	7.05
Cd	594	0.051	1.53 (0.07)	0.0477	13.8
Ce	1048	0.09	2.12 (0.09)	0.0865	7.61
Co	1768	0.152	3.75 (0.16)	0.146	4.52
Cr	2123	0.183	4.6 (0.20)	0.174	3.78
Cs	302.8	0.026	0.5 (0.02)	0.025	26
Cu	1356	0.117	3.1 (0.13)	0.111	5.93
Fe	1812	0.156			
Ga	302.8	0.026	1.336 (0.06)	0.0214	30.8
Ge	1213	0.105	7.7 (0.33)	0.0711	9.26
Hg	234	0.02	0.55 (0.02)	0.019	34.6
I_2	387	0.033	3.77 (0.16)	0.0146	44.9
In	430	0.037	0.78 (0.03)	0.0358	18.4
K	336.5	0.029	0.571 (0.025)	0.0281	23.4
La	1153	0.099	2.5 (0.11)	0.0945	6.97
Li	453	0.039	0.7 (0.03)	0.0381	17.3
Mg	923	0.08	2.1 (0.09)	0.0754	8.73
Mn	1517	0.131	(3.2) (0.14)	0.125	5.28
Mo	2873	0.248	6.6 (0.29)	0.234	2.81
N_2	63	0.005	0.172 (0.007)	0.005	131
Na	370.8	0.032	0.63 (0.027)	0.031	21.2
Ni	1728	0.149	4.22 (0.183)	0.14	4.69
O_2	44	0.004	0.106 (0.005)	0.004	184
P_4	317	0.027	0.60 (0.026)	0.026	25
Pb	600	0.052	1.15 (0.05)	0.0497	13.2
Pd	1825	0.157	4.0 (0.17)	0.15	4.4
Pr	1203	0.104	2.7 (0.12)	0.0984	6.69

(continued)

Table 5.2 (continued)

Element	Melting point, T_m, K	Thermal energy, kT_m, eV	Latent heat, $h\nu$, kcal/mol (eV)	Fluctuation of energy, Γ, eV	Lifetime, τ, $\times 10^{-15}$ s
Pt	2042	0.176	5.2 (0.23)	0.164	4
Rb	312	0.027	0.525 (0.02)	0.0261	2.52
S	392	0.034	0.3 (0.01)	0.0336	19.6
Sb_2	903.5	0.078	9.5 (0.41)	0.0293	22.5
Sc	1673	0.144	3.6 (0.15)	0.137	4.79
Se_6	493	0.042	9.0 (0.39)	0.0039	168
Si	1703	0.147	11.1 (0.48)	0.0968	6.8
Sn	505	0.044	1.69 (0.07)	0.0387	17
Sr	1043	0.09	2.1 (0.09)	0.0861	7.64
Ta	3253	0.28	5.9 (0.26)	0.271	2.43
Te	723	0.062	8.36 (0.36)	0.0202	32.6
Ti	1933	0.167	4.5 (0.20)	0.157	4.18
Tl	577	0.05	1.03 (0.05)	0.0481	13.7
U	1403	0.121	3.0 (0.13)	0.115	5.71
W	3653	0.315	8.4 (0.37)	0.298	2.21
Zn	692.5	0.06	1.74 (0.08)	0.0558	11.8
Zr	2133	0.184	4.6 (0.20)	0.175	3.76

On the other hand, boiling is also understood as follows. When the liquid metal vaporizes, if the metallic atoms have sufficient thermal energy, they will jump out of the liquid phase into the gas phase. It is considered that the energy fluctuation increases as the temperature of the liquid metal rises, and as a result, the lifetime of the metallic atoms remaining in the liquid state becomes shorter. Then, when the mean lifetime becomes equal to the characteristic period of thermal oscillation of the atom, 1.5×10^{-13} s, a kind of resonance occurs, and the vaporization proceeds like an avalanche and seems to develop into a boiling as a cooperative phenomenon.

The abovementioned explanations could be reasonably understood for considering the following facts about, for example, water. Water is vaporized under one atmosphere even at near room temperature, and water molecules individually jump out of the liquid phase to the gas phase, and then, on the other hand, the vaporization as a collective behavior, that is, as a cooperative and avalanche-like phenomenon, occurs at 100 °C. By comparing this fact about water, the reason why the average lifetime becomes equal to the characteristic period of electrons (3.3×10^{-14} s) and that of thermal oscillation of atoms (1.5×10^{-13} s) in melting and boiling, respectively, can be understood. These constant lifetimes are the typical examples of a time characteristic of the evolution of the statistical distribution discussed by Messiah (1975). These are considered to be the resonance conditions for melting and boiling to proceed at once like the avalanche as a cooperative phenomenon. Second, when comparing the variation in the lifetime of melting with the variation in the lifetime

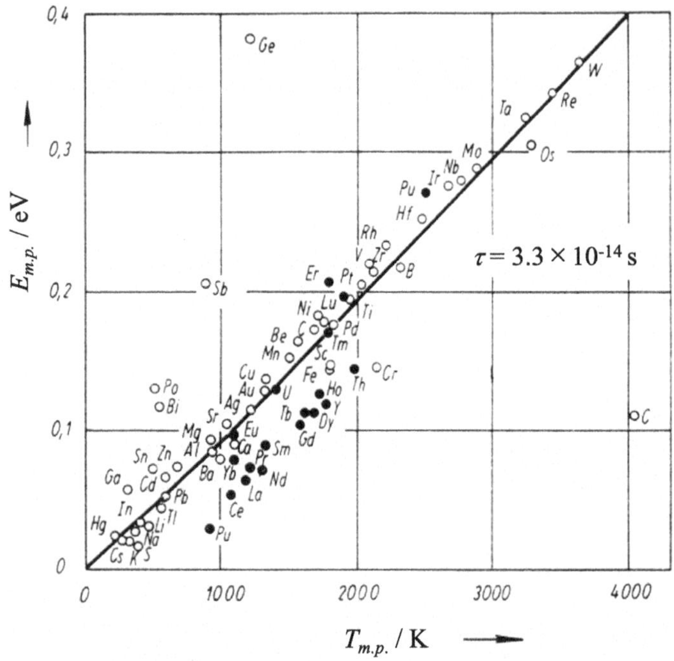

Fig. 5.2 Correlation between $E_{\text{m.p.}}$ ($h\nu$: latent heat) and $T_{\text{m.p.}}$ (melting point) [2]: The latent heat-melting point curve in this figure is given by $\tau = 3.3 \times 10^{-14}\,\text{s} = \frac{h}{2\pi E}\left\{\exp\left(\frac{E}{2kT}\right) - \exp\left(-\frac{E}{2kT}\right)\right\}$ where E is the latent heat and T is melting temperature

of boiling, we can see that the latter is smaller than the former. The reason why this is so because the binding energy varies greatly depending on the type of element, and hence the nature of the bonding, while the thermal energy does not change as much as the binding energy even if the type of element is changed. From the above mentioned, it can be understood that the interpretation of the thermal activation based on the uncertainty principle gives a fairly satisfactory explanation for the melting and boiling of metals.

Now, for the convenience of application to various phenomena, the results of numerical calculations between lifetime (τ), energy ($h\nu$), and temperature (T) are shown in Fig. 5.4 and Table 5.4.

5.3 Generalization of Johnson-Mehl Equation and Application

The phase transformation equation formulated by Johnson and Mehl [5] and Avrami [6–8] is expressed as

Table 5.3 Analysis of boiling of metals; *Databook of Metals* [3] and *Metallurgical Thermochemistry* [4]

Element	Boiling point, T_b K	Thermal energy, kT_b eV	Latent heat, $h\nu$ kcal/mol (eV)	Fluctuation of energy, Γ eV	Lifetime, τ, $\times 10^{-13}$ s
Ag	2473	0.213	60.0 (2.61)	0.00571	1.15
Al	2773	0.239	69.6 (3.03)	0.00534	1.23
Au	3223	0.278	82.0 (3.57)	0.00577	1.14
Be	2673	0.23	73.8 (3.21)	0.00302	2.18
Bi	1883	0.158	41.1 (1.79)	0.00619	1.06
Ca	1693	0.146	41.0 (1.78)	0.00399	1.65
Cd	1039	0.089	23.9 (1.04)	0.0031	2.12
Cr	2773	0.239	86.1 (3.74)	0.00149	4.41
Cs	973	0.084	15.9 (0.69)	0.0113	0.584
Cu	2843	0.245	73.3 (3.19)	0.00474	1.39
Fe	3343	0.288	81.3 (3.53)	0.0077	0.854
Ga	2523	0.217	61.4 (2.67)	0.00575	1.14
Ge	3143	0.271	78.3 (3.40)	0.00639	1.03
Hg	630	0.054	14.13 (0.61)	0.00221	2.97
I_2	456	0.039	9.96 (0.43)	0.00181	3.64
In	2348	0.202	55.5 (2.41)	0.00624	1.05
K	1052	0.091	18.9 (0.82)	0.0089	0.739
La	2973	0.256	80.0 (3.48)	0.00391	1.69
Li	1602	0.138	35.3 (1.53)	0.006	1.1
Mg	1378	0.119	30.5 (1.33)	0.00492	1.34
Mn	2368	0.204	53.7 (2.33)	0.00772	8.52
Mo	5823	0.502	121.0 (5.26)	0.0278	0.236
N_2	77	0.007	1.333 (0.06)	0.000733	8.98
Na	3183	0.100	23.7 (1.03)	0.00583	1.13
Ni	3183	0.274	89.4 (3.89)	0.00324	2.03
O_2	90	0.008	1.63 (0.071)	0.00073	9.02
P_4	553	0.048	12.4 (0.54)	0.00187	3.52
Pb	2013	0.173	42.5 (1.85)	0.00894	0.736
Pt	4373	0.377	112.0 (4.87)	0.0076	0.866
Rb	953	0.082	18.1 (0.787)	0.00653	1.01
S	898	0.077	25.4 (1.104)	0.000881	7.47
Sb_2	1948	0.168	39.4 (1.713)	0.0104	0.632
Se_6	968	0.083	21.5 (0.934)	0.00346	1.9
Si	2873	0.248	72.5 (3.15)	0.00544	1.21
Sn	3023	0.26	64.7 (2.813)	0.0127	0.518

(continued)

Table 5.3 (continued)

Element	Boiling point, T_b K	Thermal energy, kT_b eV	Latent heat, $h\nu$ kcal/mol (eV)	Fluctuation of energy, Γ eV	Lifetime, τ, $\times 10^{-13}$ s
Sr	1658	0.143	33.6 (1.46)	0.00881	0.747
Te	1263	0.109	25.7 (1.117)	0.0066	0.998
Tl	1773	0.149	39.7 (1.726)	0.00533	1.23
W	5673	0.489	183.0 (7.957)	0.00232	2.83
Zn	1180	0.102	27.3 (1.187)	0.00346	1.9
Zr	5023	0.433	128.0 (5.565)	0.00898	0.733

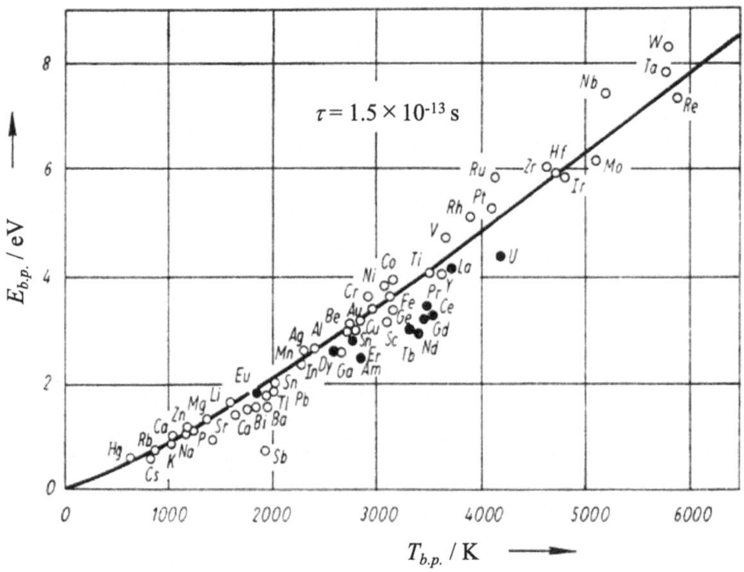

Fig. 5.3 Correlation between $E_{\text{b.p.}}$ ($h\nu$: latent heat) and $T_{\text{b.p.}}$ (boiling point) [2]: The latent heat-boiling point curve in this figure is given by $\tau = 1.5 \times 10^{-13}s = \frac{h}{2\pi E}\{\exp(\frac{E}{2kT}) - \exp(-\frac{E}{2kT})\}$ where E is the latent heat and T is boiling temperature

$$y = \exp\left[-\left(\frac{t}{\tau}\right)^n\right] \tag{5.17}$$

In Eq. (5.17), the fraction y of the old phase is given by N/N_0 (N and N_0 are the number of atoms in the old phase at times $t = t$ and $t = 0$, respectively). This equation is rewritten as

$$N = N_0\exp\left[-\left(\frac{t}{\tau}\right)^n\right] \tag{5.18}$$

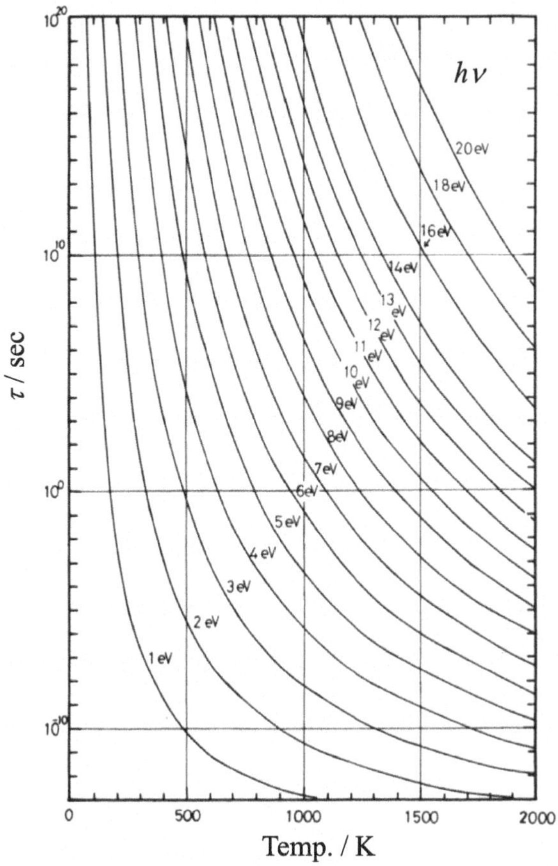

Fig. 5.4 Contour curves of energy in lifetime and temperature diagram

In this section, we will clarify the physical meaning of this expression and try to generalize it and apply it to various types of phase transformations [9, 10].

(Case-A) Decomposition of Old Phase is Rate-Determining

Phase transformation is generally a complex process consisting of several elementary processes, but first consider the case where the old phase is quite stable, and its decomposition is a rate-determining process. As described in Sect. 4.2.2, assuming that the lifetime of the old phase depends only on the energy fluctuation Γ, it is considered that one atom in the old phase is statistically transferred to the new phase with a mean lifetime τ. This process can be statistically treated as follows. If the time Δt is sufficiently short, the probability q that an atom transitions from an unstable state of old phase to a stable state of new phase is proportional to Δt, as

$$q = \lambda \cdot \Delta t \tag{5.19}$$

Table 5.4 Energy fluctuations Γ and lifetime τ in various phenomena

T (K)	kT (eV)	Specific heat capacity		
		$T_\theta = 100$ K $h\nu = 8.6165 \times 10^{-3}$ eV	$T_\theta = 300$ K $h\nu = 2.5850 \times 10^{-3}$ eV	$T_\theta = 1320$ K $h\nu = 1.1374 \times 10^{-3}$ eV
5	4.31×10^{-4}	$\Gamma = 3.91 \times 10^{-7}$ eV $\tau = 2.20 \times 10^{-9}$ s	$\Gamma = 2.42 \times 10^{-15}$ eV $\tau = 2.72 \times 10^{-1}$ s	$\Gamma = 1 \times 10^{-59}$ eV $\tau = 1.23 \times 10^{43}$ s
10	8.62×10^{-4}	$\Gamma = 5.81 \times 10^{-5}$ eV $\tau = 1.13 \times 10^{-11}$ s	$\Gamma = 8 \times 10^{-9}$ eV $\tau = 8.3 \times 10^{-8}$ s	$\Gamma = 2.47 \times 10^{-30}$ eV $\tau = 2.67 \times 10^{14}$ s
50	4.31×10^{-3}	$\Gamma = 3.67 \times 10^{-3}$ eV $\tau = 1.80 \times 10^{-13}$ s	$\Gamma = 1.29 \times 10^{-3}$ eV $\tau = 5.10 \times 10^{-13}$ s	$\Gamma = 2.10 \times 10^{-7}$ eV $\tau = 3.0 \times 10^{-9}$ s
100	8.62×10^{-3}	$\Gamma = 8.27 \times 10^{-3}$ eV $\tau = 7.96 \times 10^{-14}$ s	$\Gamma = 6.07 \times 10^{-3}$ eV $\tau = 1.08 \times 10^{-13}$ s	$\Gamma = 1.55 \times 10^{-4}$ eV $\tau = 4.25 \times 10^{-12}$ s
200	1.72×10^{-2}	$\Gamma = 1.71 \times 10^{-2}$ eV $\tau = 3.86 \times 10^{-14}$ s	$\Gamma = 1.57 \times 10^{-2}$ eV $\tau = 4.19 \times 10^{-14}$ s	$\Gamma = 4.20 \times 10^{-3}$ eV $\tau = 1.57 \times 10^{-13}$ s
500	4.31×10^{-2}	$\Gamma = 4.30 \times 10^{-2}$ eV $\tau = 1.53 \times 10^{-14}$ s	$\Gamma = 4.24 \times 10^{-2}$ eV $\tau = 1.55 \times 10^{-14}$ s	$\Gamma = 3.27 \times 10^{-2}$ eV $\tau = 2.10 \times 10^{-14}$ s
1000	8.62×10^{-2}	$\Gamma = 8.61 \times 10^{-2}$ eV $\tau = 7.64 \times 10^{-15}$ s	$\Gamma = 8.58 \times 10^{-2}$ eV $\tau = 7.67 \times 10^{-15}$ s	$\Gamma = 8.02 \times 10^{-2}$ eV $\tau = 8.20 \times 10^{-15}$ s
3000	2.58×10^{-1}	$\Gamma = 2.58 \times 10^{-1}$ eV $\tau = 2.55 \times 10^{-15}$ s	$\Gamma = 2.58 \times 10^{-1}$ eV $\tau = 2.55 \times 10^{-15}$ s	$\Gamma = 2.56 \times 10^{-1}$ eV $\tau = 2.57 \times 10^{-15}$ s
5000	4.31×10^{-1}	$\Gamma = 4.31 \times 10^{-1}$ eV $\tau = 6.53 \times 10^{-15}$ s	$\Gamma = 4.31 \times 10^{-1}$ eV $\tau = 1.53 \times 10^{-15}$ s	$\Gamma = 4.31 \times 10^{-1}$ eV $\tau = 1.53 \times 10^{-15}$ s
10,000	8.62×10^{-1}	$\Gamma = 8.62 \times 10^{-1}$ eV $\tau = 7.67 \times 10^{-15}$ s	$\Gamma = 8.62 \times 10^{-1}$ eV $\tau = 7.64 \times 10^{-16}$ s	$\Gamma = 8.62 \times 10^{-1}$ eV $\tau = 7.64 \times 10^{-16}$ s

(continued)

Table 5.4 (continued)

T (K)	kT (eV)	Diffusion	Diffusion precipitation	Graphitization	
		$h\nu = 1$ eV	$h\nu = 5$ eV	$h\nu = 10$ eV	$h\nu = 15$ eV
5	4.31×10^{-4}				
10	8.62×10^{-4}				
50	4.31×10^{-3}	$\Gamma = 5.19 \times 10^{-51}$ eV $\tau = 1.66 \times 10^{35}$ s			
100	8.62×10^{-3}	$\Gamma = 6.29 \times 10^{-26}$ eV $\tau = 1.05 \times 10^{10}$ s			
200	1.72×10^{-2}	$\Gamma = 2.51 \times 10^{-13}$ eV $\tau = 2.62 \times 10^{-3}$ s	$\Gamma = 6.53 \times 10^{-63}$ eV $\tau = 1.32 \times 10^{47}$ s		
500	4.31×10^{-2}	$\Gamma = 9.12 \times 10^{-6}$ eV $\tau = 7.22 \times 10^{-11}$ s	$\Gamma = 3.15 \times 10^{-26}$ eV $\tau = 2.09 \times 10^{9}$ s	$\Gamma = 5.19 \times 10^{-50}$ eV $\tau = 1.66 \times 10^{34}$ s	$\Gamma = 4.90 \times 10^{-75}$ eV $\tau = 1.76 \times 10^{59}$ s
1000	8.62×10^{-2}	$\Gamma = 3.02 \times 10^{-3}$ eV $\tau = 2.18 \times 10^{-13}$ s	$\Gamma = 1.25 \times 10^{-12}$ eV $\tau = 5.25 \times 10^{-4}$ s	$\Gamma = 6.29 \times 10^{-25}$ eV $\tau = 1.05 \times 10^{9}$ s	$\Gamma = 2.36 \times 10^{-27}$ eV $\tau = 2.78 \times 10^{21}$ s
3000	2.58×10^{-1}	$\Gamma = 1.48 \times 10^{-1}$ eV $\tau = 4.46 \times 10^{-15}$ s	$\Gamma = 3.15 \times 10^{-4}$ eV $\tau = 2.09 \times 10^{-12}$ s	$\Gamma = 4.0 \times 10^{-8}$ eV $\tau = 1.7 \times 10^{-8}$ s	$\Gamma = 3.76 \times 10^{-12}$ eV $\tau = 1.75 \times 10^{-4}$ s
5000	4.31×10^{-1}	$\Gamma = 3.47 \times 10^{-1}$ eV $\tau = 1.89 \times 10^{-15}$ s	$\Gamma = 1.51 \times 10^{-2}$ eV $\tau = 4.36 \times 10^{-14}$ s	$\Gamma = 9.12 \times 10^{-5}$ eV $\tau = 7.22 \times 10^{-12}$ s	$\Gamma = 4.13 \times 10^{-7}$ eV $\tau = 2.09 \times 10^{-9}$ s
10,000	8.62×10^{-1}	$\Gamma = 8.15 \times 10^{-1}$ eV $\tau = 8.07 \times 10^{-16}$ s	$\Gamma = 2.76 \times 10^{-1}$ eV $\tau = 2.39 \times 10^{-15}$ s	$\Gamma = 3.02 \times 10^{-2}$ eV $\tau = 2.18 \times 10^{-14}$ s	$\Gamma = 2.49 \times 10^{-3}$ eV $\tau = 2.64 \times 10^{-13}$ s

This assumption implies that the probability of an atom transitioning from the old phase to the new phase in unit time is independent of time, i.e., independent of the history of the atom, and that the decomposition products of the old phase are identical for any atom. The probability p that an atom is still alive after a time of only Δt, i.e., the atom of interest remains in the old phase, is

$$p = 1 - \lambda \cdot \Delta t \tag{5.20}$$

The probability that an atom remains in the old phase at time $t = k \cdot \Delta t$ can be written down as a composite event consisting of staying from 0 to Δt, then from Δt to $2\Delta t$, ..., and so on. Then, the probability to be sought is expressed as the product:

$$(1 - \lambda \cdot \Delta t)^k = (1 - \lambda \cdot \Delta t)^{\frac{t}{\Delta t}} \tag{5.21}$$

This is rewritten as

$$\left[(1 - \lambda\Delta t)^{-\frac{1}{\lambda\Delta t}}\right]^{-\lambda t} \tag{5.22}$$

When Δt approaches 0, we can obtain

$$p = \lim_{\Delta t \to 0}\left[(1 - \lambda\Delta t)^{-\frac{1}{\lambda\Delta t}}\right]^{-\lambda t} = e^{-\lambda t} \tag{5.23}$$

The probability of one atom remaining in the old phase at time t is thus obtained as $\exp[-\lambda t]$, and then if there are N_0 atoms present at the instant $t = 0$, only the number of atoms shown in the following equation on average will remain in the old phase at time t without transitioning:

$$N = N_0 e^{-\lambda t} \tag{5.24}$$

As is clear from the discussion in Sect. 4.2.2, if the mean lifetime of the atoms remaining in the old phase is τ, then $\lambda = 1/\tau$, and therefore, Eq. (5.24) becomes

$$N = N_0 e^{-\left(\frac{t}{\tau}\right)} \tag{5.25}$$

Since such phenomenon as the transition of atoms from the old phase to the new phase has a statistical character, it can be concluded that a law expressed in Eq. (5.25) is strictly speaking only satisfied when N is very large. If N is not very large, deviations (variation) should be observed, as similarly with any statistical phenomenon. Thus, if the decomposition of the old phase is the rate-determining process in the phase transformation, we can obtain an expression corresponding to $n = 1$ in Eq. (5.18), where t corresponds to the mean lifetime of the atoms remaining in the old phase.

(Case-B) Growth of a Certain Number of Precipitates Under Diffusion Control

Next, let us consider a case where diffusion is rate-determining. For example, such a phenomenon is the case that when an isolated precipitate having a radius R and the concentration of a solute C_β grows in an infinitely large homogeneous solid solution of an initial composition C_I, the diffusion of solute atoms from the area of solid solution to the interface of the precipitate controls the rate of the overall reaction. Figure 5.5 shows the concentration profile in this case. The diffusion flux J is given by the following equation, Fick's first law:

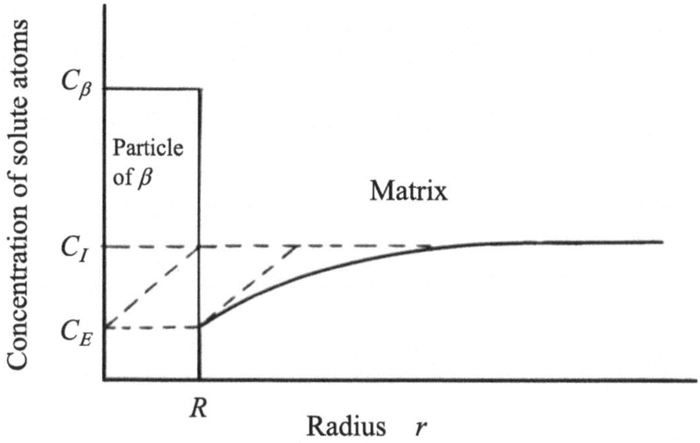

Fig. 5.5 Concentration of solute atoms in and around a growing precipitate

$$J = -D\left(\frac{\partial C}{\partial r}\right)_t \tag{5.26}$$

When the precipitate/matrix interface advances by dR during the time dt, the material balance of solute atoms is, based on the law of conservation of matter, expressed as

$$(C_\beta - C_E)(4\pi R^2)\mathrm{d}R = 4\pi R^2 D\left(\frac{\partial C}{\partial r}\right)_{r=R}\mathrm{d}t \tag{5.27}$$

The left side represents the amount of solute atoms acquired by the precipitate, and the right side represents the amount of solute atoms flowing in through the interface. From Eq. (5.27), the advancing rate of the interface dR/dt is given by

$$(C_\beta - C_E)\left(\frac{\mathrm{d}R}{\mathrm{d}t}\right) = D\left(\frac{\partial C}{\partial r}\right)_{r=R} \tag{5.28}$$

In order to find the growth rate of the precipitate particles by solving Eq. (5.28), $(\partial C/\partial r)_{r=R}$ needs to be specifically given. When the degree of supersaturation is small, $(\partial C/\partial r)_{r=R}$ can be easily obtained. In other words, the depletion of solute atoms occurs over a very large distance compared to the radius of the precipitates, while the rate of increase in R is considered to be very small. In this case, therefore, the steady-state approximation may be applied to solve the diffusion Eq. (5.26). The steady-state transfer of solute atoms toward the precipitate can be expressed using Eq. (5.26) as

$$\dot{n} = 4\pi r^2 D\left(\frac{\partial C}{\partial r}\right)_{r=r} = \text{constant} \tag{5.29}$$

Let us integrate Eq. (5.29) by assuming that D does not depend on C and the solute concentrations at R and R_2 are C_E and C_2, respectively. We obtain

$$\dot{n} = 4\pi D \frac{C_2 - C_E}{\left(\frac{1}{R} - \frac{1}{R_2}\right)} \tag{5.30}$$

From Eqs. (5.29) and (5.30), $(\partial C/\partial r)_{r=r}$ can be expressed as

$$\left(\frac{\partial C}{\partial r}\right)_{r=r} = \frac{C_2 - C_E}{\left(\frac{1}{R} - \frac{1}{R_2}\right)} \cdot \frac{1}{r^2} \tag{5.31}$$

If R_2 is infinite and $C_2 = C_I$, the gradient at the interface is

$$\left(\frac{\partial C}{\partial r}\right)_{r=R} = \frac{C_I - C_E}{R} \tag{5.32}$$

When Eqs. (5.28) and (5.32) are combined, we obtain

$$R\left(\frac{dR}{dt}\right) = D\left(\frac{C_I - C_E}{C_\beta - C_E}\right) \tag{5.33}$$

or

$$R^2 = 2D\left(\frac{C_I - C_E}{C_\beta - C_E}\right) \cdot t \tag{5.34}$$

From this equation, the time dependence of the growth rate is expressed as

$$\frac{dR}{dt} = \sqrt{D/2}\left(\frac{C_I - C_E}{C_\beta - C_E}\right)^{1/2} \cdot t^{-\frac{1}{2}} \tag{5.35}$$

Assuming that N precipitates per unit volume are generated randomly at $t = 0$ and the average radius of all precipitates becomes R at time t, then the number of solute atoms removed from the unit volume of solid solution is

$$N \cdot \frac{4}{3}\pi R^3 (C_\beta - C_E) \tag{5.36}$$

Since the total number of solute atoms that should precipitate is $(C_I - C_E)$, the ratio of transformation x is given by

$$x = N \cdot \frac{4}{3}\pi R^3 \left(\frac{C_\beta - C_E}{C_I - C_E}\right) \tag{5.37}$$

By combining Eqs. (5.34) and (5.37), the fraction of precipitation x is

$$x = \frac{8\sqrt{2}}{3}\pi N D^{3/2}\left(\frac{C_I - C_E}{C_\beta - C_E}\right)^{1/2} t^{3/2}$$
$$\cong 1 - \exp\left[-\frac{8\sqrt{2}}{3}\pi N D^{3/2}\left(\frac{C_I - C_E}{C_\beta - C_E}\right)^{1/2} t^{3/2}\right] \tag{5.38}$$

or

$$y \equiv 1 - x = \exp\left[-\frac{8\sqrt{2}}{3}\pi N D^{3/2}\left(\frac{C_I - C_E}{C_\beta - C_E}\right)^{1/2} t^{3/2}\right] = \exp\left[-\left(\frac{t}{\tau_{\text{eff}}}\right)^{3/2}\right] \tag{5.39}$$

where

$$\tau_{\text{eff}} = \left(\frac{1}{8\sqrt{2}N}\right)^{2/3}\left(\frac{C_\beta - C_E}{C_I - C_E}\right)^{1/3} \cdot \frac{1}{D} \tag{5.40}$$

D is given by Eq. (5.11) and has such temperature dependence as

$$D = \frac{\pi \nu a^2}{\left\{\exp\left(\frac{h\nu/2}{kT}\right) - \exp\left(-\frac{h\nu/2}{kT}\right)\right\}} \tag{5.41}$$

From the above consideration, it can be seen that when a certain number of spherical precipitates are grown at the same rate under diffusion control, the time exponent of Johnson-Mehl Equation (5.17) becomes $n = 3/2$. Further, the effective lifetime τ_{eff} of such a complex reaction is given by a complicated equation such as Eq. (5.40).

(Case-C) Constant Nucleation Rate of Precipitate and Diffusion-Controlled Growth

As an example, we consider the case where the number of precipitated particles per unit volume increases at a constant rate I, and the growth rate is controlled by diffusion. At time t, the radius R of the spherical precipitate particles nucleated at time $t_1 (0 < t_1 < t)$ is obtained from Eq. (5.34) as follows.

$$R^2 = 2D\left(\frac{C_I - C_E}{C_\beta - C_E}\right)(t - t_1) \tag{5.42}$$

Assuming that the volume of this particle is v, the growth rate of volume at time t is

$$\frac{dv}{dt} = 4\pi R^2 \frac{dR}{dt} = 4\sqrt{2}\pi D^{3/2}\left(\frac{C_I - C_E}{C_\beta - C_E}\right)^{3/2} \cdot (t - t_1)^{1/2} \tag{5.43}$$

In almost all precipitation reactions, the volume occupied by the precipitates is negligibly small compared to that of the matrix. Therefore, the number of precipitates formed between time t_1 and $(t_1 + \mathrm{d}t_1)$ is $I\mathrm{d}t_1$, and the increase rate in the volume of this group of precipitates at time t is $I\mathrm{d}t_1 \cdot (\mathrm{d}v/\mathrm{d}t)$. The increase rate of the volume V of all particles formed from $t = 0$ to $t = t$ is

$$\frac{\mathrm{d}V}{\mathrm{d}t} = 4\sqrt{2}\pi D^{3/2}\left(\frac{C_I - C_E}{C_\beta - C_E}\right)^{3/2} \cdot I \int_0^t (t - t_1)^{1/2}\mathrm{d}t_1$$
$$= \frac{8\sqrt{2}}{3}\pi D^{3/2}\left(\frac{C_I - C_E}{C_\beta - C_E}\right)^{3/2} \cdot I \cdot t^{3/2} \tag{5.44}$$

In order to convert the volume of the precipitate into the fraction of precipitation, Eq. (5.44) must be multiplied by $(C - C_E)/(C_I - C_E)$. Then, by integrating Eq. (5.44), the fraction of precipitation x becomes

$$x = \frac{16\sqrt{2}}{15}\pi D^{3/2}\left(\frac{C_I - C_E}{C_\beta - C_E}\right)^{1/2} \cdot I \cdot t^{5/2}$$
$$\cong 1 - \exp\left[-\frac{16\sqrt{2}}{15}\pi D^{3/2}\left(\frac{C_I - C_E}{C_\beta - C_E}\right)^{1/2} \cdot I \cdot t^{5/2}\right] \tag{5.45}$$

or

$$y \equiv 1 - x = \exp\left[-\frac{16\sqrt{2}}{15}\pi D^{3/2}\left(\frac{C_I - C_E}{C_\beta - C_E}\right)^{1/2} \cdot I \cdot t^{5/2}\right]$$
$$= \exp\left[-\left(\frac{t}{\tau_{\mathrm{eff}}}\right)^{5/2}\right] \tag{5.46}$$

where

$$\tau_{\mathrm{eff}} = \left(\frac{15}{16\sqrt{2}\pi I}\right)^{2/5} \cdot \left(\frac{C_\beta - C_E}{C_I - C_E}\right)^{1/5} \cdot \frac{1}{D^{3/5}} \tag{5.47}$$

D in this equation is the diffusion coefficient as already given by Eq. (5.11):

$$D = \frac{\pi \nu a^2}{\left\{\exp\left(\frac{h\nu/2}{kT}\right) - \exp\left(-\frac{h\nu/2}{kT}\right)\right\}} \tag{5.48}$$

Equation (5.46) is the Johnson-Mehl equation with time exponent of $n = 5/2$. The effective lifetime of this reaction is given by Eq. (5.47).

(Case-D) Eutectoid Transformations and Transformations in Pure Solid Metals

It is known that the linear rate of growth is constant for crystals growing during a polymorphic transformation in pure metals or products in eutectoid domains. Such transformations differ from the abovementioned precipitations in the following two points:

(i) Interference between growing domains takes the form of direct impingement. Thus, the growth rate of a domain is constant until it impinges on another domain and the rate becomes zero.
(ii) The volume of the growing phase is not negligible in comparison with that of matrix.

Consider unit volume of parent, and let x_v be the volume of the product formed at time t (i.e., x_v is the fraction of the volume that has completed the transformation). The rate at which the product-parent interface moves in the direction perpendicular to it is denoted by G. For simplicity, the shape of the growing product phase is assumed to be spherical. The overall rate in volume increase of product is expressed by

$$\frac{dx_v}{dt} = GA \tag{5.49}$$

where A is the total area of the product-parent interface; the product is growing freely in the sample.

In order to evaluate A as a function of x_v and t, Johnson-Mehl [5] and Avrami [6–8] introduced the concepts of extended volume V_x and extended surface A_x. These are defined as the volume and surface area, respectively, of the product that would be present if all domains were to grow without impingement and if new domains were continuously nucleated throughout the specimen including regions where transformation has already terminated. In other words, V_x is the sum of the volume of the actual transformed domain and the volume of the "phantom" domains that would have nucleated in the transformed regions if they had been untransformed [9].

V_x and A_x defined in this way are related by an equation similar to Eq. (5.49):

$$\frac{dV_x}{dt} = GA_x \tag{5.50}$$

In case of a random distribution of nuclei, due to the occurrence of random nucleation, the fraction of A_x that is not within a domain and hence is "free" interface area is equal to the untransformed volume fraction, $1 - x_v$. A becomes

$$A = (1 - x_v)A_x \tag{5.51}$$

Combining Eqs. (5.49), (5.50), and (5.51) gives

$$\frac{\mathrm{d}x_v}{\mathrm{d}V_x} = 1 - x_v \tag{5.52}$$

Integrating with separation of variables, we obtain

$$-\ln(1 - x_v) = V_x$$

or

$$x_v = 1 - \mathrm{e}^{-V_x} \tag{5.53}$$

The transformed volume at time t of domains nucleated at time t_1 is $4\pi G^3(t - t_1)^3/3$. If I is the rate of nucleation per unit volume of matrix, then the number of domains formed between time t_1 and $t_1 + \mathrm{d}t_1$ is

$$\mathrm{d}n = I(1 - x_v)\mathrm{d}t_1 \tag{5.54}$$

The number of the "phantom" nuclei that would have formed in the transformed volume is $Ix_v\mathrm{d}t_1$, and thus, the total number including the "phantom" one is

$$\mathrm{d}n_x = I\mathrm{d}t_1 \tag{5.55}$$

Hence, the extended volume at time t is

$$V_x = \frac{4}{3}\pi G^3 \int_0^t (t - t_1)^3 \cdot I \cdot \mathrm{d}t_1 \tag{5.56}$$

Combination of Eqs. (5.53) and (5.56) gives the equation for x_v as a function of t. For example, if N nuclei are formed at $t = 0$ and none thereafter, t_1 and $\mathrm{d}t_1$ are zero and

$$V_x = \frac{4}{3}\pi (Gt)^3 N \tag{5.57}$$

and

$$x_v = 1 - \exp\left[-\frac{4}{3}\pi N G^3 t^3\right] \tag{5.58}$$

or

$$y_v \equiv 1 - x_v = \exp\left[-\frac{4}{3}\pi N G^3 t^3\right] = \exp\left[-\left(\frac{t}{\tau_{\mathrm{eff}}}\right)^3\right] \tag{5.59}$$

where

$$\tau_{\text{eff}} = \left(\frac{3}{4\pi N G^3} \right)^{1/3} \tag{5.60}$$

Equation (5.59) is the Johnson-Mehl equation with exponent $n = 3$, and the corresponding effective lifetime τ_{eff} is given by Eq. (5.60).

If the rate of nucleation I is constant, V_x becomes

$$V_x = \frac{1}{3}\pi I G^3 t^4 \tag{5.61}$$

and

$$x_v = 1 - \exp\left[-\frac{1}{3}\pi I G^3 t^4 \right] \tag{5.62}$$

or

$$y_v \equiv 1 - x_v = \exp\left[-\frac{1}{3}\pi I G^3 t^4 \right] = \exp\left[-\left(\frac{t}{\tau_{\text{eff}}} \right)^4 \right] \tag{5.63}$$

where

$$\tau_{\text{eff}} = \left(\frac{3}{\pi I G^3} \right)^{1/4} \tag{5.64}$$

Equation (5.63) is the Johnson-Mehl equation with time exponent $n = 4$, and Eq. (5.64) is its effective lifetime.

(Case-E) Exponential Decrease of Nucleation Rate and Diffusion-Controlling Growth Rate

If the growth rate of new-phase particles I is given by

$$I = \frac{\mathrm{d}N}{\mathrm{d}t} = \frac{N_0}{\tau_2} \exp\left(-\frac{t}{\tau_2} \right) \tag{5.65}$$

The number of particles N at time t is

$$N = N_0 \left[1 - \exp\left(-\frac{t}{\tau_2} \right) \right] \tag{5.66}$$

On the other hand, when the interface of new and old phases moves under diffusion control, the radius R of the spherical new-phase particle at time t is given by Eq. (5.34) as

$$R^2 = 2D\left(\frac{C_I - C_E}{C_\beta - C_E} \right) \cdot t \tag{5.67}$$

From Eqs. (5.65) and (5.67), the volume of the new phase formed by time t is

$$\begin{aligned} V &= \frac{4}{3}\pi(2D)^{3/2}\left(\frac{C_I - C_E}{C_\beta - C_E}\right)^{3/2}\frac{N_0}{\tau_2}\int_0^t (t-s)^{3/2}\exp\left(-\frac{t}{\tau_2}\right)\mathrm{d}s \\ &= \frac{4}{3}\pi(2D)^{3/2}\left(\frac{C_I - C_E}{C_\beta - C_E}\right)^{3/2} N_0\frac{2}{5}t^{3/2} \\ &\quad \left[\frac{t}{\tau_2} - \frac{2}{7}\left(\frac{t}{\tau_2}\right)^2 + \frac{53}{98}\left(\frac{t}{\tau_2}\right)^3 + \frac{77}{6237}\left(\frac{t}{\tau_2}\right)^4 + \cdots\right] \\ &\cong \frac{8}{15}\pi(2D)^{3/2}\left(\frac{C_I - C_E}{C_\beta - C_E}\right)^{3/2} \cdot N_0 \cdot t^{3/2}\left[1 - \exp\left(-\frac{t}{\tau_2}\right)\right] \end{aligned} \tag{5.68}$$

Multiplying Eq. (5.68) by $(C_\beta - C_E)/(C_I - C_E)$, the precipitation ratio x is obtained as

$$x = 1 - \exp\left[-\frac{16\sqrt{2}}{15}\pi D^{3/2}\left(\frac{C_I - C_E}{C_\beta - C_E}\right)^{1/2} \cdot N_0 \cdot t^{3/2}\left\{1 - \exp\left(-\frac{t}{\tau_2}\right)\right\}\right] \tag{5.69}$$

or the ratio y of the old phase is

$$y \equiv 1 - x = \exp\left[-\left(\frac{t}{\tau_1}\right)^{3/2}\left\{1 - \exp\left(-\frac{t}{\tau_2}\right)\right\}\right] \tag{5.70}$$

where

$$\tau_1 = \left(\frac{15}{16\sqrt{2}\pi N_0}\right)^{2/3}\left(\frac{C_\beta - C_E}{C_I - C_E}\right)^{1/3} \cdot \frac{1}{D} \tag{5.71}$$

From Eq. (5.11),

$$D = \frac{\pi \nu a^2}{\left\{\exp\left(\frac{h\nu/2}{kT}\right) - \exp\left(-\frac{h\nu/2}{kT}\right)\right\}} \tag{5.72}$$

Equation (5.70) is a generalized form of Eq. (5.39) or (5.46).

(Case-F) Nucleation Rate of New Phase Decreases Exponentially and Linear Rate of Growth is Constant

By assumption, the nucleation rate of new phases, I, is given by

$$I = \frac{\mathrm{d}N}{\mathrm{d}t} = \frac{N_0}{\tau_2}\exp\left(-\frac{t}{\tau_2}\right) \tag{5.73}$$

In this case, the number of particles N at time t is

$$N = N_0\left[1 - \exp\left(-\frac{t}{\tau_2}\right)\right] \tag{5.74}$$

On the other hand, since the linear rate of growth of the new phase is constant, its radius R is

$$R = Gt \tag{5.75}$$

From Eqs. (5.72) and (5.74), the equation corresponding to Eq. (5.56) is obtained as

$$\begin{aligned} V_x &= \frac{4}{3}\pi G^3 \cdot \frac{N_0}{\tau_2}\int_0^t (t-s)^3 \cdot \exp\left(-\frac{t}{\tau_2}\right)\mathrm{d}s \\ &= 8\pi N_0 G^3 \tau_2^3\left\{\mathrm{e}^{-\left(\frac{t}{\tau_2}\right)} - 1 + \frac{t}{\tau_2} - \frac{1}{2}\left(\frac{t}{\tau_2}\right)^2 + \frac{1}{6}\left(\frac{t}{\tau_2}\right)^3\right\} \end{aligned} \tag{5.76}$$

In case of $\tau_2 \gg t$, from Eq. (5.76), we obtain

$$V_x \risingdotseq \frac{1}{3}\pi N_0 G^3 t^3\left\{1 - \exp\left(-\frac{t}{\tau_2}\right)\right\} \tag{5.77}$$

Therefore,

$$x_v = 1 - \mathrm{e}^{-V_x} = 1 - \exp\left[-\frac{1}{3}\pi N_0 G^3 t^3\left\{1 - \exp\left(-\frac{t}{\tau_2}\right)\right\}\right] \tag{5.78}$$

or

$$y_v \equiv 1 - x_v = \exp\left[-\left(\frac{t}{\tau_1}\right)^3\left\{1 - \exp\left(-\frac{t}{\tau_2}\right)\right\}\right] \tag{5.79}$$

where

$$\tau_1 = \left(\frac{3}{\pi N_0 G^3}\right)^{1/3} \tag{5.80}$$

In case of $\tau_2 \ll t$, we obtain

$$V_x \doteqdot \frac{4}{3}\pi N_0 G^3 t^3 \tag{5.81}$$

Therefore,

$$x_v = 1 - \mathrm{e}^{-V_x} = 1 - \exp\left(-\frac{4}{3}\pi N_0 G^3 t^3\right) \tag{5.82}$$

or

$$y_v \equiv 1 - x_v = \exp\left(-\frac{4}{3}\pi N_0 G^3 t^3\right) = \exp\left[-\left(\frac{t}{\tau_1}\right)^3\right] \tag{5.83}$$

where

$$\tau_1 = \left(\frac{3}{4\pi N_0 G^3}\right)^{1/3} \tag{5.84}$$

Equation (5.83) naturally agrees with Eq. (5.59). Equation (5.79) has a generalized form of these equations.

(Case-G) Nucleation Rate of New Phase Increases Exponentially and Linear Rate of Growth is Constant

Since the number of new-phase particles increases exponentially, we can write down

$$N = N_0 \exp\left(\frac{t}{\tau_2}\right), \quad I = \frac{\mathrm{d}N}{\mathrm{d}t} = \frac{N_0}{\tau_2}\exp\left(\frac{t}{\tau_2}\right) \tag{5.85}$$

On the other hand, as for the growth rate, the linear rate of growth is constant. That is, the radius R of the new phase is given as

$$R = Gt \tag{5.86}$$

In this case, an equation corresponding to Eq. (5.56) is obtained using Eqs. (5.85) and (5.86) as

$$\begin{aligned} V_x &= \frac{4}{3}\pi G^3 \cdot \frac{N_0}{\tau_2}\int_0^t (t-s)^3 \exp\left(\frac{t}{\tau_2}\right)\mathrm{d}s \\ &= \frac{4}{3}\pi N_0 G^3 (\tau_2)^3 \left\{ 2\mathrm{e}^{\left(\frac{t}{\tau_2}\right)}\left[\left(\frac{t}{\tau_2}\right)^3 - 3\left(\frac{t}{\tau_2}\right)^2 + 6\left(\frac{t}{\tau_2}\right) - 3\right] \right. \\ &\quad \left. + \left[\left(\frac{t}{\tau_2}\right)^3 - 3\left(\frac{t}{\tau_2}\right)^2 + 6\right]\right\} \end{aligned} \tag{5.87}$$

The effect of the exponential increase in the number of new-phase particles becomes significant at $t \gg \tau_2$. In the case of $t \gg \tau_2$, Eq. (5.87) is written as

$$V_x = \frac{8}{3}\pi N_0 G^3 t^3 \exp\left(\frac{t}{\tau_2}\right) \tag{5.88}$$

Therefore,

$$x_v = 1 - e^{-V_x} = 1 - \exp\left[-\frac{8}{3}\pi N_0 G^3 t^3 \exp\left(\frac{t}{\tau_2}\right)\right] \tag{5.89}$$

or

$$y_v \equiv 1 - x_v = \exp\left[-\frac{8}{3}\pi N_0 G^3 t^3 \exp\left(\frac{t}{\tau_2}\right)\right] = \exp\left[-\left(\frac{t}{\tau_1}\right)^3 \exp\left(\frac{t}{\tau_2}\right)\right] \tag{5.90}$$

where

$$\tau_1 = \left(\frac{3}{8\pi N_0 G^3}\right)^{1/3} \tag{5.91}$$

Equation (5.90) is equivalent to the generalized Eq. (5.59). When $t \ll \tau_2$, Eq. (5.90) has the same form of Eq. (5.59).

From the above examples, it can be seen that the time exponent n and the effective lifetime τ vary with the transformation mechanism in the Johnson-Mehl equation for phase transformation, which is shown in the following equation.

$$y = \exp\left[-\left(\frac{t}{\tau}\right)^n\right] \tag{5.92}$$

Further thinking about Eqs. (5.70) and (5.79) yields

$$y = \exp\left[-\left(\frac{t}{\tau_1}\right)^{n_1} \cdot \left\{1 - \exp\left(-\frac{t}{\tau_2}\right)\right\}^{n_2}\right] \tag{5.93}$$

Furthermore, in the same way, extending Eq. (5.90), we obtain

$$y = \exp\left[-\left(\frac{t}{\tau_1}\right)^{n_1} \cdot \exp\left(\frac{t}{\tau_2}\right)^{n_2}\right] \tag{5.94}$$

The results are summarized in Table 5.5.

The above discussion is for cases where only one type of reaction occurs. In reality, however, not only one type of reaction occurs, and an apparent single phenomenon is generally considered to be composed of multiple types of reactions. The relationship

Table 5.5 Various types of phase transformations and their rate equations

Model		Rate equation
(a) *Polymorphic transformation, discontinuous precipitation, eutectoid formation, growth controlled by the interface process*		
(1)	Decomposition of the old phase is the rate-determining case	$y = \exp\left[-\left(\frac{t}{\tau}\right)\right]$
(2)	Nucleation of a new phase at grain boundaries	$y = \exp\left[-\left(\frac{t}{\tau}\right)\right]$
(3)	Nucleation of new phase at the tip of particles	$y = \exp\left[-\left(\frac{t}{\tau}\right)^2\right]$
(4)	Nucleation rate of a new phase is zero (the number of particles is constant)	$y = \exp\left[-\left(\frac{t}{\tau}\right)^3\right]$
(5)	Nucleation rate of a new phase decreases exponentially	$y = \exp\left[-\left(\frac{t}{\tau_1}\right)^3\left\{1 - \exp\left(-\frac{t}{\tau_2}\right)\right\}\right]$
(6)	Nucleation rate of a new phase is constant	$y = \exp\left[-\left(\frac{t}{\tau}\right)^4\right]$
(7)	Nucleation rate of a new phase increases exponentially	$y = \exp\left[-\left(\frac{t}{\tau_1}\right)^3 \cdot \exp\left(\frac{t}{\tau_2}\right)\right]$
(b) *Growth controlled by diffusion*		
(1)	Precipitation on dislocations	$y = \exp\left[-\left(\frac{t}{\tau}\right)^{1/2}\right]$
(2)	Thickness of very large disk-like particles increases	$y = \exp\left[-\left(\frac{t}{\tau}\right)^{1/2}\right]$
(3)	Radial thickness of columnar or needle-like particles increases	$y = \exp\left[-\left(\frac{t}{\tau}\right)\right]$
(4)	Growth of needle or plate-like particles (the length of particle is smaller than the distance between particles)	$y = \exp\left[-\left(\frac{t}{\tau}\right)\right]$
(5)	Growth of phases that have considerable volume from the beginning	$y = \exp\left[-\left(\frac{t}{\tau}\right)^{1\sim 3/2}\right]$
(6)	Arbitrarily shaped phases growing from fine particles (The number of particles is constant)	$y = \exp\left[-\left(\frac{t}{\tau}\right)^{3/2}\right]$
(7)	Arbitrarily shaped phases growing from fine particles (nucleation rate decreases exponentially)	$y = \exp\left[-\left(\frac{t}{\tau_1}\right)^{3/2}\left\{1 - \exp\left(-\frac{t}{\tau_2}\right)\right\}\right]$

(continued)

Table 5.5 (continued)

Model		Rate equation
(8)	Arbitrarily shaped phases growing from fine particles (nucleation rate is constant)	$y = \exp\left[-\left(\frac{t}{\tau}\right)^{5/2}\right]$
(9)	Arbitrarily shaped phases growing from fine particles (nucleation rate increases exponentially)	$y = \exp\left[-\left(\frac{t}{\tau_1}\right)^3 \cdot \exp\left(\frac{t}{\tau_2}\right)\right]$

between multiple reactions that apparently constitute a single phenomenon can be roughly divided into "parallel relation" and "serial relation".

When n reactions in total occur in parallel, the overall reaction can be expressed as the sum of the individual reactions as

$$Y = \sum_{i=1}^{n} C_i y_i \tag{5.95}$$

$$\sum_{i=1}^{n} C_i = 1 \tag{5.96}$$

Each y_i is given by the rate equation as shown in Table 5.5, according to the model of that reaction.

Next, we will consider the case where multiple reactions start in series, that is, starting from some substance A_1, and A_1 changes to A_2, and then A_2 changes to A_3, …

$$A_1 \to A_2 \to A_3 \to A_4 \to \cdots \tag{5.97}$$

As an example of such a sequential reaction, we consider that only one ring in the chain of reactions is separated from the others, and an amount of the material exponentially decreases, as in Eq. (5.25). Even when the rate equation of the individual reactions is such simple, the sequential reactions generally follow fairly complicated kinetics. We will suppose that the number of atoms in the materials in the series (5.97) is N_1, N_2, N_3, … at time t, and the number of atoms in the first material (the ancestor of the series) is $N_1(0)$ at the instant $t = 0$. For the reaction of $A_1 \to A_2$, Eq. (5.98) holds:

$$\frac{dN_1}{dt} = -\frac{1}{\tau_1} N_1 \tag{5.98}$$

Integrating Eq. (5.98) gives

$$N_1 = N_1(0)\exp\left(-\frac{t}{\tau_1}\right) \tag{5.99}$$

In the series $A_1 \rightarrow A_2 \rightarrow A_3$ consisting of two rings, Eq. (5.100) is satisfied in addition to Eq. (5.98).

$$\frac{dN_2}{dt} = \frac{1}{\tau_1}N_1 - \frac{1}{\tau_2}N_2 \tag{5.100}$$

This equation describes the fact that A_2 atoms arise from A_1 at a rate of N_1/τ_1 and A_2 atoms themselves become A_3 at a rate of N_2/τ_2. Equation (5.100) is rewritten using Eq. (5.99) as

$$\frac{dN_2}{dt} + \frac{1}{\tau_2}N_2 = \frac{1}{\tau_1}N_1(0)\exp\left(-\frac{t}{\tau_1}\right) \tag{5.101}$$

A particular solution of this first-order linear inhomogeneous equation is found as

$$N_2 = N_1(0)\left\{C_1\exp\left(-\frac{t}{\tau_1}\right) + C_2\exp\left(-\frac{t}{\tau_2}\right)\right\} \tag{5.102}$$

By adding the term $N_2(0)\exp[-(t/\tau_2)]$ to this particular solution, we obtain the general solution as

$$N_2(t) = N_2(0)\exp\left(-\frac{t}{\tau_2}\right) + N_1(0)\left\{C_1\exp\left(-\frac{t}{\tau_1}\right) + C_2\exp\left(-\frac{t}{\tau_2}\right)\right\} \tag{5.103}$$

Since the initial condition at $t = 0$ is clearly $N_2(0) = 0$, we are only interested in the particular solution, Eq. (5.102). Substituting Eq. (5.102) into Eq. (5.101), a simple calculation leads

$$C_1 = \frac{\tau_2}{(\tau_1 - \tau_2)} \tag{5.104}$$

Therefore,

$$N_2(t) = N_1(0)\left\{\frac{\tau_2}{(\tau_1 - \tau_2)}\exp\left(-\frac{t}{\tau_1}\right) + C_2\exp\left(-\frac{t}{\tau_2}\right)\right\} \tag{5.105}$$

Using the initial condition of $N_2(0) = 0$ at $t = 0$, Eq. (5.105) gives us

$$C_2 = -\frac{\tau_2}{(\tau_1 - \tau_2)} \tag{5.106}$$

Finally,

$$N_2(t) = \frac{N_1(0)\cdot\tau_2}{(\tau_1-\tau_2)}\left\{\exp\left(-\frac{t}{\tau_1}\right)-\exp\left(-\frac{t}{\tau_2}\right)\right\} \tag{5.107}$$

If the series of reactions consists of three rings, that is, $A_1 \rightarrow A_2 \rightarrow A_3 \rightarrow A_4$, we add

$$\frac{dN_3}{dt} = \frac{1}{\tau_2}N_2 - \frac{1}{\tau_3}N_3 \tag{5.108}$$

By substituting the right side of Eq. (5.107) into N_2 in Eq. (5.108) and moving N_3/τ_3 to the left side, we can obtain

$$\frac{dN_3}{dt} + \frac{1}{\tau_3}N_3 = \frac{N_1(0)}{(\tau_1-\tau_2)}\left\{\exp\left(-\frac{t}{\tau_1}\right)-\exp\left(-\frac{t}{\tau_2}\right)\right\} \tag{5.109}$$

The solution is

$$N_3(t) = N_1(0)\left\{C_1\exp\left(-\frac{t}{\tau_1}\right)+C_2\exp\left(-\frac{t}{\tau_2}\right)+C_3\exp\left(-\frac{t}{\tau_3}\right)\right\} \tag{5.110}$$

The constants C_1, C_2, and C_3 are determined in a similar way of abovementioned procedure as

$$\begin{aligned} C_1 &= \frac{\tau_1\cdot\tau_3}{(\tau_1-\tau_2)(\tau_1-\tau_3)} \\ C_2 &= \frac{\tau_2\cdot\tau_3}{(\tau_2-\tau_1)(\tau_2-\tau_3)} \\ C_3 &= \frac{\tau_3\cdot\tau_3}{(\tau_3-\tau_1)(\tau_3-\tau_2)} \end{aligned} \tag{5.111}$$

Therefore, Eq. (5.110) becomes

$$\begin{aligned} N_3(t) = N_1(0)\Bigg\{&\frac{\tau_1\cdot\tau_3}{(\tau_1-\tau_2)(\tau_1-\tau_3)}\exp\left(-\frac{t}{\tau_1}\right)+\frac{\tau_2\cdot\tau_3}{(\tau_2-\tau_1)(\tau_2-\tau_3)}\exp\left(-\frac{t}{\tau_2}\right) \\ &+\frac{\tau_3\cdot\tau_3}{(\tau_3-\tau_1)(\tau_3-\tau_2)}\exp\left(-\frac{t}{\tau_3}\right)\Bigg\} \end{aligned} \tag{5.112}$$

In the case of a series consisting of four rings: $A_1 \rightarrow A_2 \rightarrow A_3 \rightarrow A_4 \rightarrow A_5$, the quantity of the fourth product is expressed by Eq. (5.113) as a function of time.

$$N_4(t) = N_1(0)\left\{C_1\exp\left(-\frac{t}{\tau_1}\right)+C_2\exp\left(-\frac{t}{\tau_2}\right)+C_3\exp\left(-\frac{t}{\tau_3}\right)+C_4\exp\left(-\frac{t}{\tau_4}\right)\right\} \tag{5.113}$$

where

$$
\begin{aligned}
C_1 &= \frac{\tau_1^2 \cdot \tau_4}{(\tau_1 - \tau_2)(\tau_1 - \tau_3)(\tau_1 - \tau_4)} \\
C_2 &= \frac{\tau_2^2 \cdot \tau_4}{(\tau_2 - \tau_1)(\tau_2 - \tau_3)(\tau_2 - \tau_4)} \\
C_3 &= \frac{\tau_3^2 \cdot \tau_4}{(\tau_3 - \tau_1)(\tau_3 - \tau_2)(\tau_3 - \tau_4)} \\
C_4 &= \frac{\tau_4^2 \cdot \tau_4}{(\tau_4 - \tau_1)(\tau_4 - \tau_2)(\tau_4 - \tau_3)}
\end{aligned} \tag{5.114}
$$

In general, in the case of a series consisting of n rings: In the case of $A_1 \rightarrow A_2 \rightarrow \cdots \rightarrow A_n \rightarrow A_{n+1}$, the quantity of the n-th product is expressed as a function of time by

$$
\begin{aligned}
N_n(t) = N_1(0) \Bigg\{ & C_1 \exp\left(-\frac{t}{\tau_1}\right) + C_2 \exp\left(-\frac{t}{\tau_2}\right) + \cdots + \\
& C_{n-1} \exp\left(-\frac{t}{\tau_{n-1}}\right) + C_n \exp\left(-\frac{t}{\tau_n}\right) \Bigg\}
\end{aligned} \tag{5.115}
$$

where

$$
\begin{aligned}
C_1 &= \frac{(\tau_1)^{n-2} \cdot \tau_n}{(\tau_1 - \tau_2)(\tau_1 - \tau_3) \cdots (\tau_1 - \tau_n)} \\
C_2 &= \frac{(\tau_2)^{n-2} \cdot \tau_n}{(\tau_2 - \tau_1)(\tau_2 - \tau_3) \cdots (\tau_2 - \tau_n)} \\
&\cdots \\
C_{n-1} &= \frac{(\tau_{n-1})^{n-2} \cdot \tau_n}{(\tau_{n-1} - \tau_1)(\tau_{n-1} - \tau_2) \cdots (\tau_{n-1} - \tau_n)} \\
C_n &= \frac{(\tau_n)^{n-2} \cdot \tau_n}{(\tau_n - \tau_1)(\tau_n - \tau_2) \cdots (\tau_n - \tau_{n-1})}
\end{aligned} \tag{5.116}
$$

In order to analyze actual concrete reactions, it is necessary to determine the rate equation that agrees well with the experimental data by taking into consideration various factors such as

(1) how many kinds of reactions they consist of,
(2) whether they proceed in parallel or serial relation,
(3) what kind of model is compatible with the reactions.

If we can obtain a rate equation that agrees with the experimental data, we can inversely know how many types of reactions the phenomenon of our concern is composed, whether those reactions progress in parallel or in series, what the mechanism works in the reaction, and so on. Actually, using the Johnson-Mehl equation (5.92) and its generalized types of Eqs. (5.93) and (5.94), we have analyzed eutectoid transformation, graphitization of cementite, precipitation from a supersaturated solid solution, etc. and achieved considerable results. Since many specific examples can be found in other books and original papers, they are omitted here.

5.4 Graphitization of Cementite by Impact Deformation

In the applications discussed above, such as diffusion, melting and boiling of metals, and applications of Johnson-Mehl equation to the phase transformation, the energy fluctuation in the uncertainty relation $\Gamma \cdot \tau \cong \hbar$ comes from increase of temperature. In contrast, the example described below is detailed because the energy fluctuations are caused by mechanical work and there are no other examples.

This experiment investigated the effect of hot impact deformation on graphitization of cementite in Fe–C binary alloy of which chemical compositions are shown in Table 5.6 [11].

Using high-purity iron and carbon, the alloy was melted in a kryptol furnace and cast in a sand mold to get white iron under atmospheric condition. The ingot was cut into disk-type specimens having 16 mm in diameter and 6 mm in thick. The specimen was heated at 1100 °C for 5 or 10 min in an electric furnace under argon atmosphere, and within 2–3 s after removing the specimen from the furnace, a 60 kg steel block was dropped from a height of 1 m onto the specimen. Such process of hot impact deformation was repeated. The experimental procedure is shown in Table 5.7.

Figures 5.6 and 5.7 show the micro-structures of the samples. As shown in Fig. 5.6a, the sample as cast has a typical micro-structure of white pig iron, and directional growth of ledeburite is observed. Figure 5.6b is the micro-structure after annealing at 1100 °C for 3 h. No graphitization was observed in the sample of this Fe–C binary alloy, while the graphitization in Fe–C–Si ternary alloys is completed by annealing at 1100 °C for 30 min. Figure 5.6c is a micro-graph of the sample of which thickness was reduced by 51% after repeating the hot impact deformation four times. Cementite has undergone a quite severe plastic deformation, and the directional alignment of cementite observed under as cast condition is completely broken. However, no graphitization is observed at this stage of hot impact deformation. Figure 5.6d is a micro-structure of the sample annealed at 1000 °C for 10 h after hot impact deformation of 49%. Although coarsening of cementite is observed, graphitization has not yet occurred. Figure 5.6e is the structure near cracks in the sample annealed at 1100 °C for 10 min after impact deformation by 33%. It is well known that the presence of defects such as dislocations or cracks in a matrix promotes

Table 5.6 Chemical compositions of a Fe–C binary alloy (mass %) [11]

C	Si	Mn	P	S	N	O	Fe
4.24	0.005	0.0002	0.006	0.013	0.0051	0.0051	Rem

Table 5.7 Experiment procedure [11]

No.	Conditions of heating and impact deformation	Photographs
1	As cast	Figure 5.6a
2	Annealing (1100 °C, 3 h)	Figure 5.6b
3	Heating (1100 °C, 5 min) → impact (21%)* → heating (1100 °C, 5 min) → impact (32%) → heating (1100 °C, 5 min) → impact (39%) → heating (1100 °C, 5 min) → impact (51%)	Figure 5.6c
4	Heating (1100 °C, 5 min) → impact (23%) → heating (1100 °C, 5 min) → impact (31%) → heating (1100 °C, 5 min) → impact (40%) → heating (1100 °C, 5 min) → impact (49%) → annealing (1000 °C, 10 h)	Figure 5.6d
5	Heating (1100 °C, 5 min) → impact (33%) → annealing (1100 °C, 10 min)	Figure 5.6e
6	Heating (1100 °C, 5 min) → impact (37%) → heating (1100 °C, 5 min) → impact (77%)	Figure 5.6f
7	Heating (1100 °C, 5 min) → impact (35%) → heating (1100 °C, 5 min) → impact (73%)	Figure 5.7

*Percentage reduction in thickness of a specimen by impact deformation

graphitization. However, the present experiment with Fe–C binary alloy shows that neither plastic deformation nor cracking of the matrix can cause graphitization.

When the degree of deformation due to the impact deformation becomes very large, significant changes appear in the micro-structure of a sample. A typical micro-structure at 77% thickness reduction is shown in Fig. 5.6f. Figure 5.7 also shows a cross-sectional photograph of a sample with 73% impact deformation and the results of analysis using X-ray micro-analyzer (EPMA). It is clear from these results that both decomposition of eutectic cementite and formation of graphite occurred. This graphitization should be considered to occur almost instantaneously. This is because the structure shown in Fig. 5.6c was observed just before the final impact deformation, and the impact deformation and subsequent cooling from 1100 °C to several hundred degrees took only a few seconds. The graphite flakes are distributed along the direction of the plastic flow caused by the deformation, which implies that the distribution of the graphite flakes is affected by the defects and the texture caused by their density and distribution.

The effects of hot impact deformation on the white iron of Fe-4.24%C binary alloy described above are summarized as follows. The effect of impact deformation is divided into two distinct stages; in the first stage where the thickness reduction does not exceed 70%, the solidified structure with the directionally oriented cementite is changed to the texture where small, crushed cementite pieces are irregularly distributed. No graphitization was observed at all even after impact deformation of 50% at 1100 °C followed by annealing at 1000 °C for 10 h. In the second stage of impact deformation, where the thickness reduction exceeds 70%, a considerable amount of graphitization occurs almost instantaneously. Graphite flakes are distributed along the direction of plastic flow due to the impact deformation (see Fig. 5.8).

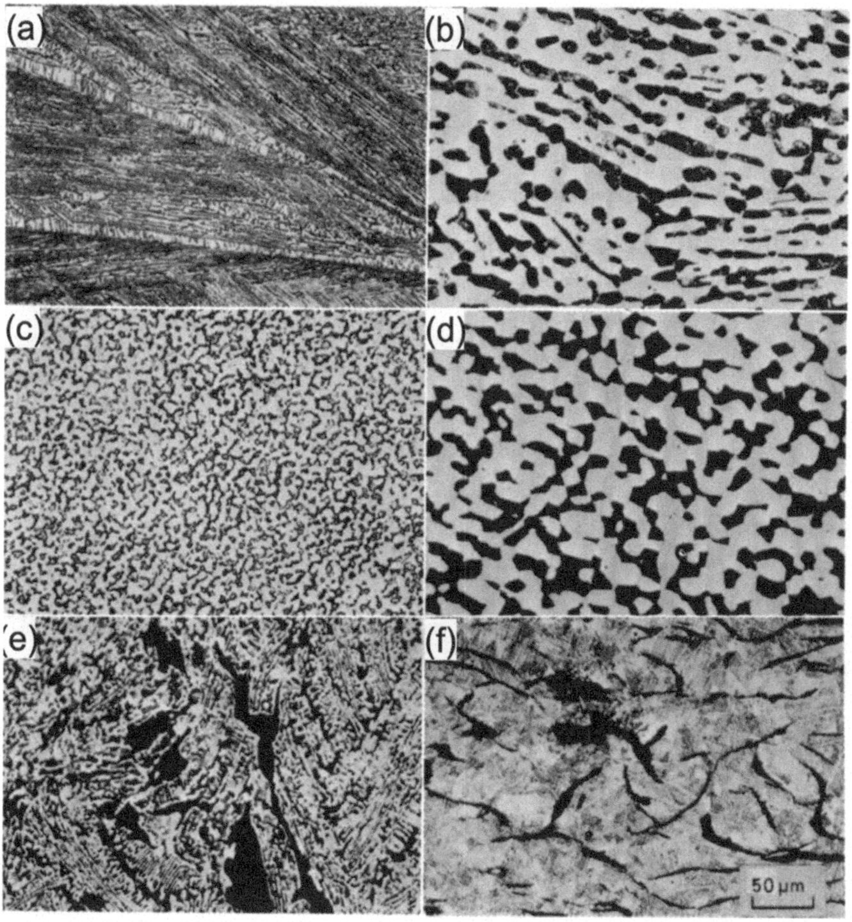

Fig. 5.6 Effect of hot impact deformation on the graphitization of cementite in Fe-4.24%C binary alloy, **a** As Cast, ledeburite (eutectic structure), **b** Annealing at 1100 °C for 3 h, **c** 51% reduction in thickness by hot impact deformation (HID) at 1100 °C, **d** 49% reduction by HID at 1100 °C and annealing at 1000 °C for 10 h, **e** 33% reduction by HID at 1100 °C and annealing at 1100 °C for 10 min, micro-structure near cracks, **a–e** no graphitization, and **f** 77% reduction by HID at 1100 °C, the occurrence of graphitization distributed along the direction of plastic flow [11]

The results of this experiment described above are extremely important for ascertaining the rate-determining process in the reaction of graphite growth and for elucidating the graphitization mechanism. Let us first consider the rate-determining process in the reaction of graphite growth. This reaction is not simple, but it consists of (1) decomposition of cementite, (2) diffusion of C atoms in an iron matrix, (3) crystallization of graphite, and (4) diffusion of Fe and Si atoms away from graphite to form space for graphite growth. Of these processes, the slowest process is the rate-determining process in the reaction of graphite growth.

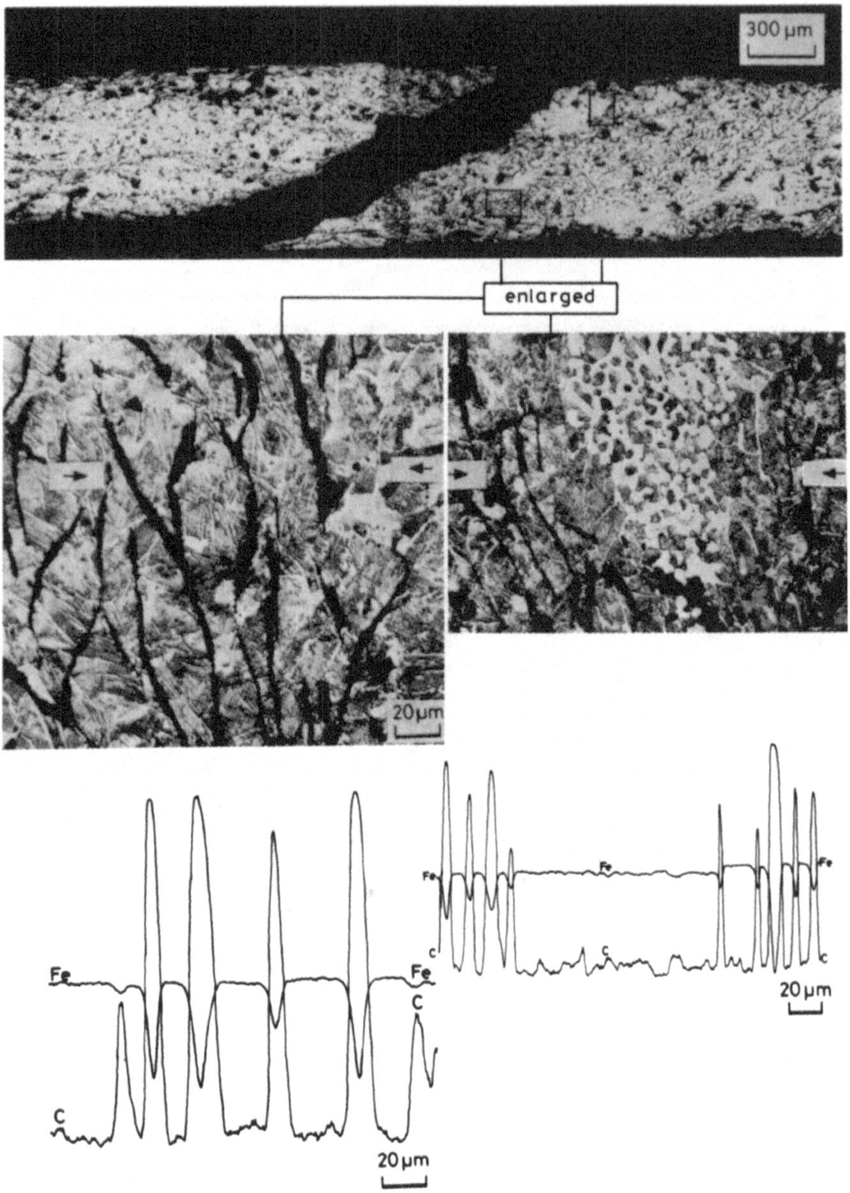

Fig. 5.7 Micro-structures and component analysis by X-ray micro-analyzer of the sample with 73% reduction of hot impact deformation at 1100 °C. Graphitization was observed in all cross sections. Component analysis was performed between two arrows [11]

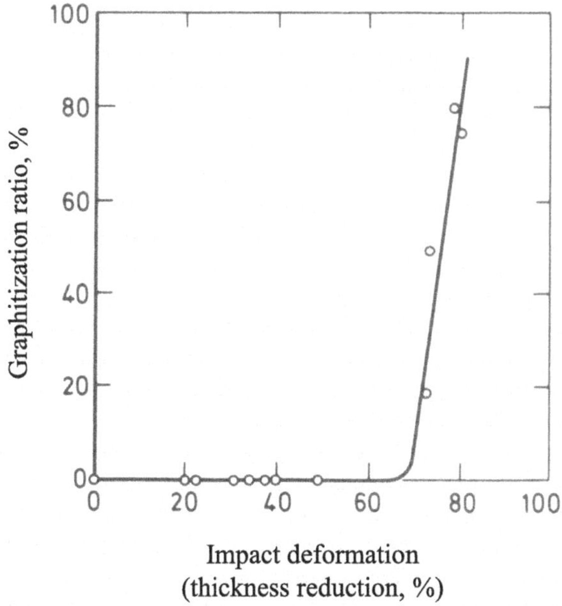

Fig. 5.8 Effect of impact deformation (thickness reduction, %) on graphitization ratio (%) in Fe-4.24%C binary alloy [11]

By comparing the results of this experiment with the diffusion data, it is possible to clarify that the diffusion of C, Fe, Si, and other elements cannot be the rate-determining process in the reaction of graphite growth. The activation energies of diffusion of Fe, Si, and other similar elements in iron are almost the same. On the other hand, that of C atom is quite small. The diffusion distance L can be roughly estimated by $L = \pi\sqrt{Dt}$ (D is diffusion coefficient, and t is diffusion time). Approximate values of the diffusion distance between Fe and C atoms are shown in Table 5.8. These estimated values at 1100 °C mean that the graphitization can considerably proceed within a few minutes. In fact, in the case of Fe–C–Si alloys, it has been reported that graphitization was completed within 30 min at 1100 °C. In the present experiment, the coarsening of cementite is possible only after considerable long-distance diffusion has occurred, but no graphitization is observed. Thus, it can be concluded that at least in the Fe–C binary system, the diffusion of C or Fe atoms will not be the rate-determining process for the reaction of graphite growth.

On the other hand, if the crystallization of graphite is the rate-determining step in the growth process, the growth rate is given by

Table 5.8 Diffusion distance at 1100 °C [11]

Element	D_0	Q	$x \cong \pi\sqrt{D_0\exp(-Q/kT)\cdot t}$ (cm)		
	$cm^2\ s^{-1}$	$kcal\ mol^{-1}$	5 s	5 min	3 h
C	0.01	27.0	5.0×10^{-3}	3.9×10^{-2}	2.3×10^{-1}
Fe	58	74.2	6.6×10^{-5}	5.1×10^{-3}	3.1×10^{-3}

$$4\pi R^2 \frac{\mathrm{d}R}{\mathrm{d}t} \cdot \rho = 4\pi R^2 \phi \tag{5.117}$$

Here, R is the radius of the graphite particles, ρ is the density of the graphite, and ϕ is the weight of C per unit surface area and unit time transported from austenite in equilibrium with graphite to graphite crystals. In this case, the growth rate is constant throughout the reaction period, and if the reaction starts at time 0, the growth curve is a straight line passing through the origin. Such results have not been obtained by other researchers, and also in the present experiment, the graphitization was not observed at all during the process of annealing at 1000 °C for 10 h but observed after the strong impact deformation. The graphitization in this case occurs almost instantaneously, which is far from the situation where the growth rate is constant throughout the reaction period. Thus, the possibility that the crystallization of graphite becomes the rate-determining process disappears.

From the above considerations, it is clear that diffusion of C, Fe, and other elements is not rate-determining process in the reaction of graphite growth, nor the crystallization of graphite. Therefore, it can be concluded that the reaction rate of graphite growth is controlled by the decomposition of cementite. As described in the experimental results, when the degree of impact deformation is not so high, no graphitization is observed at all even when the matrix is strongly plastic deformed and small cracks are generated. Therefore, sudden onset of graphitization cannot be attributed to plastic deformation or the formation of micro-cracks in the matrix. Therefore, a serious question arises as to why cementite was almost instantaneously decomposed by the strong impact deformation.

To discuss this issue, the crystal structure of cementite and the nature of the bonds in cementite must be taken into account. Determination of the crystal structure and lattice spacing of cementite has been an important issue for a long time. According to the most accurate measurement of the structure of cementite, it is an orthorhombic lattice belonging to the space group *Pnma*, with lattice constants a = 5.0896 Å, b = 6.7443 Å, c = 4.5248 Å. As shown in Fig. 5.9, there are eight iron atoms at d-position, four iron atoms at c-position, and four carbon atoms at c-position. As shown in the figure, the atomic arrangement can be regarded as a layer in which the dcd-layer and the $\overline{\mathrm{d}}\,\overline{\mathrm{c}}\,\overline{\mathrm{d}}$-layer are alternately stacked. Each carbon atom is surrounded by six iron atoms. Two of them are in the same layer as the carbon atoms, and the other four are in the d- or $\overline{\mathrm{d}}$-layers. These six iron atoms form a triangular prism as shown in Fig. 5.10. As shown in Fig. 5.9a, a set of dcd-layers (or $\overline{\mathrm{d}}\,\overline{\mathrm{c}}\,\overline{\mathrm{d}}$-layers) has a network structure in which such triangular prisms are connected to each other by iron atoms at their corners.

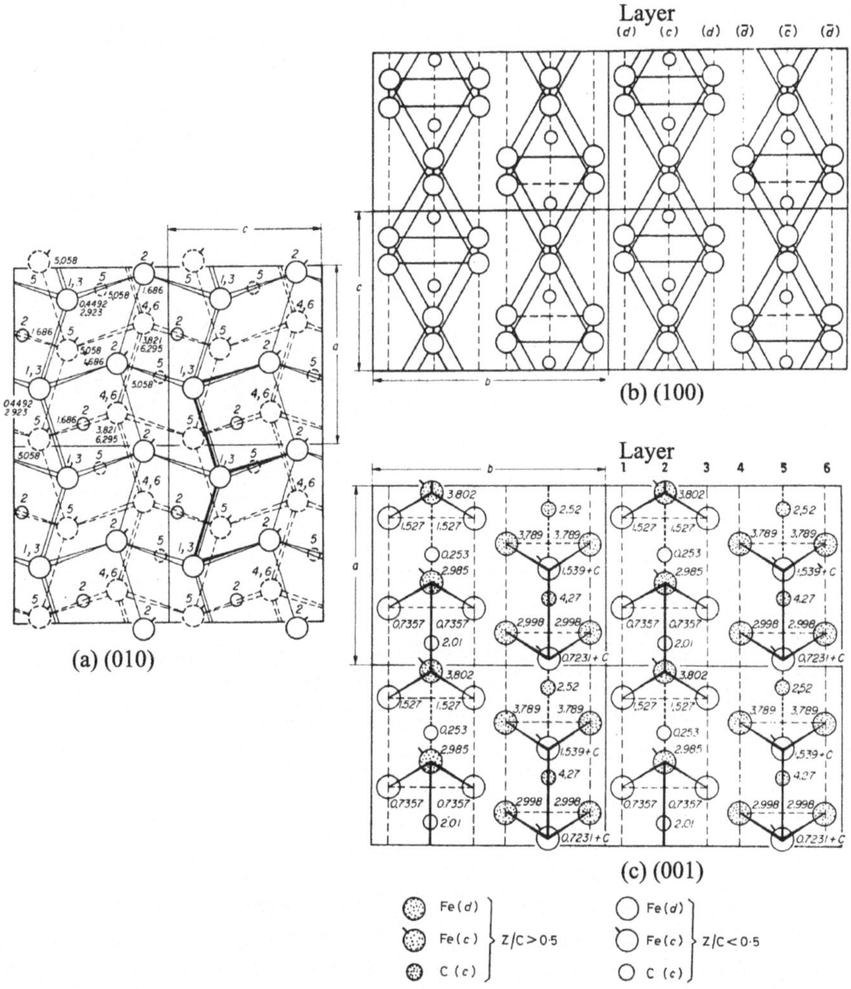

Fig. 5.9 Crystal structure of cementite [11], **a** Projection on (010) plane. Number beside each atom indicates (010) layer which it is situated; lower numbers represent planes closest to observer. **b** Projection on (100) plane. **c** Projection on (001) plane

The nature of bonding in cementite is discussed based on this crystal structure of cementite. The first notable feature of the structure is the existence of Fe–Fe bond (bonding distance 2.47–2.69 Å) and Fe–C bond (bonding distance 1.97–2.04 Å). It is suggested that Fe–C bond has covalency, while Fe–Fe bond has metallicity, and that the strength of Fe–C bond is about twice that of Fe–Fe bond, etc. Thus, we can say that the bonds in the triangular prism shown in Fig. 5.10 are covalent bond, which is stronger than the metallic bond that connects the triangular prisms together.

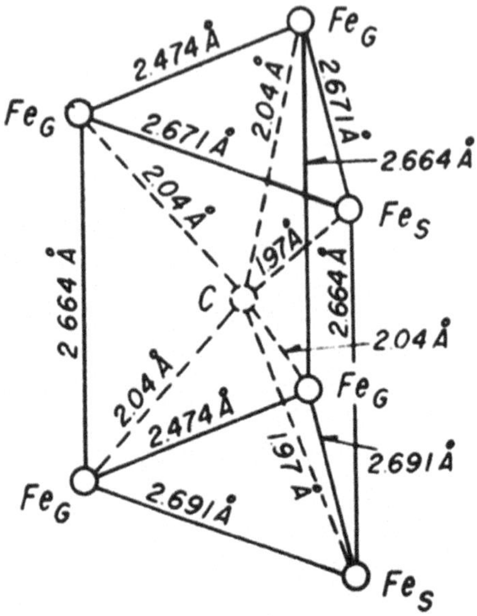

Fig. 5.10 Arrangements of Fe atoms around a C atom in cementite [11]

Why plastic deformation is possible with metals? Plastic deformation is possible in metals because the nature of bonding is metallic. In the case of structures with covalent bonding, electron clouds are significantly concentrated in certain direction, and this directional bonding cannot be maintained during deformation, and once broken, it is not easily recovered. For plastic deformation to occur, in order to retain the nearest-neighbor atomic arrangement, especially the Fe–C bonding, the disturbance should be as little as possible.

Thus, if stress is applied to a crystal of cementite, plastic deformation may be possible to occur by the relative motion of the triangular prisms, without destroying the triangular prisms themselves. Such plastic deformation is possible, for example, by sliding along the (010) plane, as can be easily seen from the layered arrangement of the triangular prisms in Fig. 5.9. Several studies support this notion. The defects existing in the cementite are generated only by changing the stacking way of the atomic layers without destroying the triangular prism itself.

It is not so easy to explain the almost instantaneous decomposition of cementite due to high-impact deformation. However, the following explanation could be possible. In the first stage of the hot impact deformation, plastic deformation of cementite crystals would be possible by the relative movement of triangular prisms, for example, by sliding along the (010) plane of the cementite, because the bond between the triangular prisms is metallic. When the degree of impact deformation becomes very large, such ultimate limits are reached, above which the covalent bonds within the unit cell of triangular prism are broken. Thus, if the deformation exceeds this limit, the cementite will collapse.

By the way, this instantaneous collapse of cementite is understood based on the uncertainty relation. Let us suppose that the potential energy when a 60 kg steel ingot is dropped from a height of 1 m is converted to the energy fluctuation in the chemical bonding of cementite. Then, the energy fluctuation per atom is

$$\Gamma = \frac{mgh}{\frac{W}{M} \cdot N_0} = \frac{60 \times 10^3(\mathrm{g}) \times 980\left(\mathrm{cm/s^2}\right) \times 100(\mathrm{cm})}{\frac{8.44(\mathrm{g})}{55.85(\mathrm{g/mol})} \cdot 6.02 \times 10^{23}\left(\mathrm{mol^{-1}}\right)} \fallingdotseq 4.0 \times 10^{-2}(\mathrm{eV}) \tag{5.118}$$

Therefore, the lifetime τ is obtained as

$$\tau = \frac{\hbar}{\Gamma} \fallingdotseq 10^{-13}(\mathrm{s}) \tag{5.119}$$

This value of the lifetime is the same order of magnitude as the period of motion of atoms. Inversely, we can make such an interpretation for the reason why the instantaneous collapse of cementite occurs due to the strong impact deformation that the conversion from the potential energy to the binding energy of the atoms proceeds efficiently and the chemical bonding in cementite is broken when τ becomes the same as the period of motion of atoms, i.e., when a kind of resonance state is allowed.

5.5 Further Applications

By the previous section, we have been developing our reaction kinetics for application to phase transformation and chemical reactions. Among them, diffusion and melting/boiling of metals are regarded as the most basic and elemental phenomena. The generalization of the Johnson-Mehl equation and its applications are for analyzing more complex phenomena. In fact, chemical reactions in alloys such as phase transformation, thermal decomposition, redox reaction can be analyzed by this generalized equation. In the analysis of these complex phenomena, it is important to consider what elementary reactions are involved and how the phenomena are combined and to use the Johnson-Mehl type equations that are appropriate for these reactions. Thus, by maintaining logical consistency between the physical content and the equation that represents it, an elucidation of the physical content of the reaction is ensured.

Although the above description deals with reactions caused by thermal activation, the graphitization of cementite by impact deformation has been described as an example of a reaction that is not caused by thermal activation. This phenomenon is of great interest because it is a phenomenon that cannot be interpreted nor does not give us any clue of solution in the conventional way of thinking, and therefore, the experimental method, results, and analysis were described in detail. Another example of transformation not caused by thermal activation is a martensitic transformation. In this case as well, the explanation can be made based on the uncertainty relation as shown in Table 3.1.

As mentioned above, regardless of whether the reaction is simple or not, and whether it is caused by thermal activation or not, our reaction kinetics based on the uncertainty principle is applicable to an extremely wide range of phase transformations and chemical reactions. The uncertainty relation holds for almost all interactions we know, from very strong to very weak interactions, as shown in Table 3.1. Therefore, as long as the phenomenon that we are interested in is caused by these interactions, in principle, our kinetic theory can be applied. However, in actual phenomena, it is necessary to consider the complicated hierarchical structure, and it is important to use an appropriate rate equation that reflects this hierarchy.

References

1. W. Hume-Rothery, *The Structure of Alloys of Iron: An Elementary Introduction* (Pergamon Press, England, 1966), Japanese edition translated by K. Hirano (Kyoritsu Shuppan, Tokyo, 1968)
2. S. Yamamoto, The time energy uncertainty principle and thermal activation. Z. Phys. Chem. **290**, 17–32 (1989). https://doi.org/10.1515/zpch-1989-27003
3. The Japan Institute of Metals and Materials, *Databook of Metals* (Maruzen, Tokyo, 1984) (in Japanese)
4. O. Kubaschewski, E. Ll. Evans, *Metallurgical Thermochemistry*, 3rd edn. (Pergamon Press, Oxford, 1967)
5. W.A. Johnson, R.F. Mehl, Reaction kinetics in processes of nucleation and growth. Trans. Am. Inst. Min. Metall. Eng. **135**, 416–458 (1939)
6. M. Avrami, Kinetics of phase change I. J. Chem. Phys. **7**, 1103–1112 (1939)
7. M. Avrami, Kinetics of phase change II. J. Chem. Phys. **8**, 212–224 (1940)
8. M. Avrami, Kinetics of phase change III. J. Chem. Phys. **9**, 117–184 (1941)
9. J. Burke, *The Kinetics of Phase Transformations in Metals* (Pergamon Press, Oxford, 1965)
10. J.W. Christian, *The Theory of Transformation in Metals and Alloys* (Pergamon Press, Oxford, 1965)
11. S. Yamamoto, Y. Kawano, N. Hattori, Y. Murakami, R. Ozaki, Influence of hot impact deformation on graphitization in white cast iron. Metal Sci. **11**, 571–577 (1977). https://doi.org/10.1179/msc.1977.11.12.571

Further Reading

The Chemical Society of Japan, *Nonequilibrium States and Relaxation Processes*, KagakuSosetsu, No. 5 (Academic Publishing Center, Tokyo, 1974) (in Japanese)

Chapter 6
Characteristics of Our Reaction Kinetics

Abstract This chapter summarizes the basic features of our reaction kinetics mainly in contrast to Eyring's absolute reaction kinetics. First, it is examined whether the theory is appropriate as a reaction kinetics that theoretically discusses time-dependent changes in phenomena, i.e., whether or not the energy of the reacting system is considered definite, whether or not the theory has a clear indication for the directionality of change, and whether or not an equilibrium is considered between the reacting system and transition state. Next, several requirements that reaction kinetics must meet as a theoretical system (e.g., the application fields of research subjects that the theory can deal with, the consistency between the hierarchy of subjects and the theory, and so on) and whether it meets the "three-step verification" as a science will be examined. Finally, the epistemological contrast between the metaphysical and dialectical worldviews represented by Aristotle and Hegel, respectively, since ancient times is presented.

Keywords Eigenvalue equation · Transition state · Criterion of truthfulness · Dialectical worldview

In this chapter, we will discuss the characteristics of our reaction kinetics developed in Chaps. 3–5 from the perspective established in Chap. 1.

6.1 Comparison with Conventional Theories

In order to compare our reaction kinetics with the conventional theories, we will first discuss some of the important issues raised in the critique of the absolute reaction kinetics and the nucleation theory in Chap. 2.

S. Yamamoto, *Reaction Kinetics Based on Time-Energy Uncertainty Principle*,
https://doi.org/10.1007/978-981-19-9673-3_6

6.1.1 Scope of Reaction Kinetics

The purpose of Eyring's absolute reaction kinetics was to theoretically calculate the activation energy E and the frequency factor A of Arrhenius Equation (2.1). Therefore, its target reactions are limited to those caused by the temperature increase. The nucleation theory deals with the process of nucleation formed by thermal fluctuations. Thus, what both theories have in common is that they focus on reactions that proceed by thermal activation. However, chemical reactions are not limited to those that occur by thermal activation. There are also reactions such as photochemical reactions. On the other hand, some nucleation theories, such as martensitic transformation, do not depend on thermal activation. No unified theory has not yet been proposed to deal with the rates of various kinds of chemical reactions and phase transformations including those of independent on thermal activation.

In contrast, our reaction kinetics based on the time–energy uncertainty principle $\Delta t \cdot \Delta E \cong \hbar$, makes it possible to know the reaction rates for various reactions that are caused not only by thermal activation but also by light, electric, or mechanical actions, as long as the magnitude of energy fluctuation, ΔE, due to these actions can be evaluated. Thus, our reaction kinetics can, in principle, make it possible to fully describe various reactions in the wide range of fields (see Sects. 2.1.1 and 2.2).

6.1.2 Assumption of Definiteness in Energy

In the absolute reaction kinetics, the eigenvalue equations are solved to construct the potential energy surface. As mentioned in Sects. 2.1.2 and 4.1.2, the use of eigenvalue equations implies that the energy is definite. The potential energy surface, by the way, is used in the consideration of activated complexes. After all, the absolute reaction kinetics assumes that the energy is definite even for transition states such as the activated state.

On the other hand, in the nucleation theory, the assumption such that the (free) energy is a single-valued function of the state is also applied to nuclei having critical sizes, and the concept of activation free energy is used (see Sect. 2.2). However, as discussed in Sects. 3.1 and 4.1, the energy becomes uncertain in the transition state. Quantitative calculation of the degree of energy uncertainty yielded the uncertainty relation. Therefore, our reaction kinetics actively takes into account the energy uncertainty in the transition state and is established as the theoretical system based on it. In our kinetics, it is no longer necessary to assume the definiteness of energy in the transition state.

6.1.3 Fundamental Laws Giving Directionality of Change

As discussed in Sect. 2.1.3, we cannot find any ideas that give the directionality of change through the methodology using the potential energy surface in the absolute reaction kinetics. The phenomena described by the wave equation in quantum mechanics or the equation of motion in classical mechanics are reversible. Therefore, the motion of particles on the potential energy surface which is described by these equations is also reversible. In fact, the sum of the potential energy and kinetic energy of a particle on a potential energy surface is constant and conserved and thus cannot explain the difference in stability between reactants and products.

On the other hand, in the thermodynamic formulation of the reaction rate in the absolute reaction kinetics and in the thermodynamic treatment of the nucleation theory, the directionality of change seems to be given by the law of entropy increase or the law of free energy decrease. However, the directionality of change involved in the law of free energy decrease has lost its meaning due to the assumption of equilibrium between the reactants and the activated complexes in the absolute reaction kinetics, and due to the assumption of equilibrium about the embryos of critical size in the nucleation theory. Even if such assumption of equilibrium was never made, thermodynamic quantities are not suitable for describing the reaction rates because they are state quantities defined for equilibrium states and do not include the concept of time.

In our reaction kinetics, the uncertainty relation $\Delta t \cdot \Delta E \cong \hbar$ plays a role of the fundamental law that indicates the directionality of change. This is because the particle of interest can remain only for a time Δt in the state which has energy fluctuation of ΔE and then the state transitions to a new state. This law gives the directionality of change in such a way that the state will change from a short-lived state to a long-lived state, just similarly as the law of entropy increase gives the directionality of change in the way that a state changes to the direction of increasing entropy. Moreover, this law of uncertainty relation includes the concept of time, which is appropriate for describing reaction rates (see Sects. 2.1.3, 2.2.2, 3.2.2, and 4.2.2).

6.1.4 Equilibrium Assumption in Transition States

In the absolute reaction kinetics, two equilibrium assumptions are made. The first is the assumption of chemical equilibrium between reactants and activated complexes, and the second is the assumption of temperature equilibrium in the transition state (see Sect. 2.1.4). In the nucleation theory, equilibrium is assumed for the distribution of embryos (see Sect. 2.2.1). We have already seen in the respective sections that these equilibrium assumptions lead to fundamental contradictions in their theoretical systems.

Our reaction kinetics does not in principle require such unreasonable equilibrium assumptions. As discussed in Sect. 4.2.1, in the interpretation of $\Delta t \cdot \Delta E \cong \hbar$, the temperature equilibrium is assumed to calculate the energy fluctuations ΔE. Therefore, there seems something to be problematic even in our reaction kinetics. However, if one considers the logic behind the construction of the reaction kinetics, we can see that the meaning of temperature equilibrium in our reaction kinetics is quite different from the meaning of the equilibrium assumption in the absolute reaction kinetics or the nucleation theory. In the absolute reaction kinetics, chemical equilibrium is assumed between reactants and activated complexes, and in the nucleation theory, equilibrium is assumed for embryos. In other words, they assume an equilibrium in the reaction itself. This introduces a contradiction into the logical structure of reaction kinetics itself, since reactions occur only because they are not in equilibrium.

On the other hand, in the treatment of thermal activation in Sect. 4.2.1, equilibrium is assumed for temperature, but not for the reaction itself. The reaction of reactants proceeds at a constant temperature because they are not in equilibrium with respect to the reaction. There is no contradiction between keeping the temperature of the reactants constant and proceeding with the reaction. So, why did temperature equilibrium in the transition state become a problem in the absolute reaction kinetics? **First**, because this assumption is incompatible with the assumption that the sum of kinetic energy (which is also thermal energy) and potential energy is constant and conserved for the motion of particles on the potential energy surface. **Second**, because it assumes a temperature to a transition state with an extremely short lifetime, that is, a transition state that lasts only for a shorter time than is required to achieve temperature equilibrium. Temperature cannot be defined without this temperature equilibrium.

These are the reasons why the equilibrium assumption in the absolute reaction kinetics and the nucleation theory leads to a logical contradiction in the construction of reaction kinetics, while the assumption of temperature equilibrium in our reaction kinetics theory does not fall into a logical contradiction.

Furthermore, it should be emphasized that the equilibrium assumption is not necessary in principle for our reaction kinetics. On the basis of the uncertainty relation, all that is needed is the fluctuation width of energy, which is independent of whether the system is in equilibrium or not. In short, all we need to know is the fluctuation width of energy, and we do not need to make any assumptions of equilibrium such as the reaction itself being in equilibrium. In fact, in the treatment of graphitization of cementite due to impact deformation in Sect. 5.4, no such equilibrium assumption was needed at all to determine the fluctuation width of energy.

6.1.5 *Problems About Hierarchical Perspective*

Lack of hierarchical perspective also leads to serious confusion and contradiction in establishing the reaction kinetics. As already clarified in Sect. 2.1.5, **the absolute reaction kinetics** results in various confusions and contradictions as follows.

1. **Absolute Reaction Kinetics**

   (1) Confusion among three different treatments for the activation energy, i.e., a treatment with potential energy surface, thermodynamic treatment, and statistical treatment. Although these are all called same "energy", they are different in character.
   (2) Automatous/superficial combination of quantum and classical mechanics everywhere.
   (3) Ignoring the hierarchical difference of substances in motion.
   (4) Ignoring the complexity and hierarchy in structure of reactions.

2. **Nucleation Theory**

   Next, there are problems related to hierarchy in **the nucleation theory**:

   (1) Usage of the macroscopic systems of logic and macroscopic concepts for describing the microscopic phenomena.
   (2) Bringing matters that become a problem and make sense only in the macro-phenomena or macro-worlds into the micro-phenomena or micro-worlds (see Sect. 2.2).

3. **Our Viewpoints**

   In **our reaction kinetics**, on the other hand, the subjects to be dealt were thoroughly examined from such hierarchical viewpoints as,

   (1) whether the object to be described is microscopic or macroscopic,
   (2) whether the energy fluctuation is quantum or classical,
   (3) whether the nature of energy in question is quantum mechanical or thermodynamical,
   (4) whether the reaction is elementary or complex, and so on.

We should be careful to use concepts and logical systems which are appropriate for the target objects and to maintain the hierarchical viewpoints both in their contents and forms as possible.

6.1.6 *Problems as Theoretical System*

Next, we will discuss the problems as a theoretical system in accordance with the items pointed out in Chap. 1.

First of all, it must be a deductive logical system based on a few well-established fundamental laws in order for the reaction kinetics to be a theoretical system.

However, the absolute reaction kinetics, as described in Sect. 2.1.6, lacks the fundamental laws that predict the directionality of change, and empirical equations are used to construct the potential surface and to formulate the specific reaction rate, which are not well established. In addition, confusion and contradictions have arisen due to irrational assumptions of equilibrium.

Even in the nucleation theory, it ultimately lacks the basic laws that give the directionality of change, as explained in Sect. 2.2. Moreover, the assumption of equilibrium in the formation process of embryos is inconsistent with its logical system. Our reaction kinetics uses the uncertainty relations as the fundamental law that gives the directionality of change and in principle does not require the equilibrium assumption. So it does not cause confusion or contradiction. The laws used in the theory are the well-established fundamental laws, the distributional laws of quantum mechanics, and the deductively derived ones from them.

Second, considering from the criteria of scientific truthfulness, the absolute reaction kinetics and the nucleation theory lack consistency between the subject and the means of description or fall into logical inconsistency due to equilibrium assumptions. As for the possibility of experimental verification, direct observation of activated complexes and critical nuclei is difficult. In our reaction kinetics, the equilibrium assumption is not made because of the importance of consistency between the object and the means of description, and the logical inconsistency caused by this is avoided. In addition, activated complexes or nuclei are not directly assumed in the sense that they are distinguished from reactants and old phases and thus are not used in any way on the formulation of rate equations. Hence, their experimental verification is not originally problematic.

6.1.7 Criteria of Theoretical Transformation

The idea of our reaction kinetics can be considered in relation to the three-step verification imposed on the transformation of scientific cognition discussed in Sect. 1.4 as follows.

First, our kinetics inherits the temperature dependence of reaction rates or phase transformation rates and the observations used to support this idea, which were derived and explained, even incompletely, in the absolute reaction kinetics and the nucleation theory. In fact, the fundamental principles of quantum mechanics and the Gibbs distribution were inherited as basic concepts, and the Arrhenius equation was derived based on them.

Second, the crucial difficulties that made the absolute reaction kinetics and the nucleation theory questionable were clarified by analyzing the characteristics of chemical reactions and the logic of their description systems. The solution of the problem was based on the results of study for the hierarchical-historical development of nature and its scientific cognition.

Third, our reaction kinetics suggests that many seemingly unrelated phenomena, such as those caused by thermal activation and those caused by other actions, can be explained in a unified manner by essentially the same principle, the uncertainty relation. It also suggests that differences in the hierarchy of matters can be distinguished and systematized by their characteristic period of motion and characteristic time duration of interaction.

Our reaction kinetics correctly inherits the results of the conventional reaction kinetics, and by adopting the completely new principles, it provides a possibility of elucidating reaction rates from a new perspective and of describing them in a unified manner.

6.2 Worldview of Our Reaction Kinetics—Dialectical Worldview

We will point out that the worldview of our reaction kinetics based on the time–energy uncertainty relation described in Chap. 3 goes beyond the metaphysical worldview and leads to the dialectical worldview. This is because the uncertainty principle underlying the development of our reaction kinetics seems to be a physical expression of the dialectical logic, as will be explained below.

The metaphysical world and the dialectical world are usually distinguished by the scope of their application: fragmented world and whole world, two-dimensional world and three-dimensional world, static world and dynamic world, and so on, respectively.

Hegel's Logic of Dialectics

Hegel proposed Hegelian logic, the so-called dialectical logic, as a logic of movement, especially of development, instead of the traditional formal logic. Hegel first criticized the conventional formal logic. The fundamental laws of formal logic are the law of identity, the law of contradiction, and the law of excluded middle. The law of identity is expressed in the form "A is A", the law of contradiction in the form "A is not non-A", and the law of excluded middle in the form "A is either B or non-B". These laws are the absolute laws of thinking, and we must follow them when we think, that can never be broken. Looking at these three laws now, we can see that they are all about eliminating contradictions from our thinking. The first "A is A" means that A cannot be non-A, and the third "A is either B or non-B" means that A cannot be B and non-B at the same time. Thus, both laws are attributed to the second law of contradiction. The principle of formal logic, therefore, is to prohibit the existence of contradiction in every event or thing.

By the way, these fundamental laws of formal logic seem to be absolute, even though they are formal. But is this really the case? Hegel criticized the law of identity and the law of contradiction:

First, such proposition that "A is A" or "A is not non-A" contradicts the general form of propositions, because propositions originally promise a distinction between subjects and predicates.

Second, while "scientific thinking" establishes the law of "self-identity" that all things have identity for themselves on the one hand, it also establishes the law of "difference" that all things have difference on the other hand. However, in these laws, contradictory predicates are attached to the same subject, "all things". This is the contradiction that the law of identity rejects.

Third, if it is asserted that all common experience adheres to the proposition, even though this proposition cannot be proven, the assertion is not true. The common experience does not say such ridiculous statements as "a planet is a planet, magnetism is magnetism, or spirit is spirit". As for the law of excluded middle, it is such a proposition of "determinate understanding" that seeks to keep contradictions at bay, but in doing so, falls into contradiction. This law asserts that something is either + A or − A. But by doing so, it implies that there is a third, the intermediate A itself; this A is *neither* + A *nor* − A *and is equally well* + A as − A. If + W means six miles to the west, and − W six miles to the east, and if + and − are sublated, then we are left with a six-mile path or space that exists with or without the opposition.

The above shows that the fundamental law of formal logic is to reject contradiction from thinking, but contradiction is never eliminated, but rather comes against the fundamental law of formal logic itself, which tries to escape from it, and makes it fall into self-contradiction.

In Hegelian logic, this contradiction is never eliminated because contradiction is the true nature of all things. It is suggested that this contradiction is what makes everything self-moving and alive. It is dialectics that attempts to grasp things not in their mere variety, but in their contradictory nature. The difference between formal logic and dialectics seems to be the difference between the trying to avoid all contradictions and the finding and pointing out contradictions in everything (the above on Hegel is interpreted by the philosopher/economist Mita [1–3] who was well known for his works on Hegel's "Logic" and Marx's "Capital").

As mentioned above, the dialectic, which is the logic of movement, positively accepts contradictions. And it is only through these contradictions that movement can be understood. The main laws of dialectics are summarized:

(Law 0) Every movement has its internal contradiction as a source of motion.
(Law 1) The law of the transformation from quantity to quality and vice versa.
(Law 2) The law of the unity and conflict of the opposites.
(Law 3) The law of the negation of the negation.

Dialectical Logic and Uncertainty Principle

Let us examine how the above dialectic is reflected in the time–energy uncertainty relation, $\Delta t \cdot \Delta E \cong \hbar$. First, where does the contradiction appear? What was the physical content of the energy fluctuation Γ? As discussed in Sect. 4.1.3, it is such that the superposition of different energies is its essence. For example, let us suppose that two different energies are denoted as E_1 and E_2, and consider the superposition of these two energies. Then, the energy of the system we are considering is E_1 and E_2 at the same time. Since E_2 is not E_1, such state of the system contains a contradiction within it. The state of the system with energy fluctuation Γ, as the manifestation of superposition of different energies, is not stable and represents a transition state with lifetime τ. This is exactly what the above Law 0 claims.

The uncertainty relation $\Gamma \cdot \tau \cong \hbar$ implies that the lifetime of a system, of which energy fluctuation Γ is zero, is infinite. In fact, the state with no fluctuations, or the state described by eigenvalue equations, is called a stationary state. As the energy fluctuation increases, the lifetime of a state decreases and the state changes to a new one. In other words, a quantitative change in energy fluctuation brings about a change of state, i.e., a qualitative change. This is what Law 1 claims. Where does the Law 2 appear? It is evident from the conclusion in Sect. 4.1.3. The static state, or the stationary state, is described as a state with a definite value in energy. But a state with finite lifetime, the transition state, is always expressed by the superposition of different energies. This means that a state of the system in motion always corresponds to a state of superposition of E and non-E, that is, it means the state of inseparable relationship of both energies of E and non-E, , which seems to be in a relationship of conflict and struggle. This is what constitutes the very content of contradiction. Where does the negation of the negation, Law 3, appear? The first negation of the law is involved in the fluctuation of energy itself, since the fluctuation of energy is caused by the superposition of energy E_1 and its negation, i.e., the non-E_1. The second negation of the law is manifested in the situation that such state having a fluctuation in energy lasts only a finite time and is negated and transformed into a new state. Thus, the energy–time uncertainty relation $\Gamma \cdot \tau \cong \hbar$ can be regarded as a physical expression of the dialectical law.

Reactions and phase transformations are a kind of movement, and therefore, in order to treat them correctly, the logic of movement must be included somewhere in the basic principles of their description. In the case of our reaction kinetics, the logic of dialectics is implied in the fundamental law of physics, which is the uncertainty relation. Thus, the worldview of our reaction kinetics is that of dialectics, and this dialectical logic seems to be indispensable for understanding movements in the world.

References

1. S. Mita, *Hegelian Logic and the Social Sciences* (Otsuki Shoten, Tokyo, 1977) (in Japanese)
2. S. Mita, *The Road to Hegel's Philosophy* (Otsuki Shoten, Tokyo, 1977) (in Japanese)
3. S. Mita, *Scientific Theory and Dialectics* (Otsuki Shoten, Tokyo, 1977) (in Japanese)

Further Reading

F. Engels, *Dialektik der Natur* (Dietz Verlag, Berlin, 1952)

Postscript

The original of this Japanese book is "New Reaction Rate Theory" written by Dr. Satoru Yamamoto and published by Showado in 1979. The translators of this book, Dr. Teruo Tanabe, Dr. Yoji Imai, Dr. Mahoto Takeda, Dr. Kenzo Hanawa, and myself, Dr. Hideo Yoshida, are all researchers engaged in advanced materials research and have been in contact with Dr. Satoru Yamamoto for many years. Based on these common points, we started working on the translation of this book around October 2019. The translators, Dr. Tanabe and Dr. Takeda, are researchers who were involved in research at universities such as Kyoto University and Yokohama National University, respectively, and Dr. Imai is a researcher who was involved in research at the National Institute of Advanced Industrial Science and Technology (AIST), while Dr. Hanawa and I are researchers who worked at research institutes of private companies, Showa Denko K. K. and UACJ Corporation, respectively.

Although our individual research fields are different, we share the common ground of materials research. All the above translators started translating this book with the hope that overseas researchers would become familiar with the new reaction kinetics proposed by Dr. Yamamoto by using it as a reference book in English.

In translation, drafting was divided by chapter. The most difficult chapters to translate were Chaps. 1 and 6, which clarified the author's philosophical position, and were translated by Dr. Tanabe, who has co-authored many books with Dr. Yamamoto and assisted by Dr. Imai; Chap. 2 by Dr. Imai; Chap. 3 by Dr. Takeda; Chap. 4 by Dr. Hanawa; Chap. 5 by Dr. Yoshida. Dr. Tanabe and Dr. Imai unified the terminology and other aspects of the translation. All chapters were cross-checked with each other through cross-reading, reading together, etc. Dr. Yoshida was in charge of external relations, such as organizing meetings and negotiating with publishers. Since face-to-face meetings and review meetings were not possible due to the COVID-19 pandemic, progress was checked through online meetings.

S. Yamamoto, *Reaction Kinetics Based on Time-Energy Uncertainty Principle*,
https://doi.org/10.1007/978-981-19-9673-3

In translating the book, we have been very careful to take into account the recent revision of terminology, etc. However, since we are currently unable to discuss with the author due to his illness, we are not sure if we have adequately conveyed the author's intentions in some difficult passages. Nevertheless, we consider ourselves jointly responsible for the translation.

I took charge of Chap. 5 because I recognized during my own research that the ideas developed in this book are very effective in solving industrial problems as described below, and I wanted foreign researchers to make use of them as well. I have also been painfully aware for many years that it is difficult to have my research papers understood because there is no English book that discusses the principles of the kinetic equation.

The kinetic equation derived by Dr. Yamamoto is very effective for the kinetic analysis of precipitation hardening, recovery, and recrystallization of aluminum alloys, which I have been studying [1]. In the conventional equations, when the incubation time is involved in the reaction, the concept of "embryo" is introduced and the incubation time is explained by the classical nucleation-growth theory, which has been followed for a long time. In contrast, Dr. Yamamoto assumed that the precipitation rate of new-phase particles naturally changes exponentially and quantitatively showed that Yamamoto's equation, which is based on the Johnson-Mehl equation with an additional term for the change in the number of particles, explains the incubation time. In other words, the term of the number of particles indicates the existence condition of the nucleation site, and it can explain the whole reaction in a unified manner without relying on the classical nucleation-growth theory.

Since then, it has become clear in various fields that Yamamoto's rate equation is valid for other reactions as well. This book also provides a detailed description of the original author's arguments and thought process. It would be the greatest pleasure for the original author and translators if the readers of this book could apply the ideas developed in this book to many more research fields, and we believe that the purpose of publishing this book has been fulfilled.

August 31, 2022

On behalf of the translators

Dr. Hideo Yoshida

ESD Laboratory

Former UACJ Corporation, R&D Center

Honorary Member of Japan Institute of Light Metals

Fellow of the Japan Institute of Metals

Reference

1. H. Yoshida, M. Asano, Recovery and Recrystallization Process in Commercially Pure Aluminum: The role of Dissolved Impurities and Analysis by a New Kinetics Theory, in *Recrystallization - types, technics and applications*, ed. by K. Huang (NOVA, 2020), pp. 167–259.

The manufacturer's authorised representative in the EU is Springer Nature Customer Service Centre GmbH, Europaplatz 3, 69115 Heidelberg, Germany. If you have any concerns regarding our products, please contact ProductSafety@springernature.com

Printed and bound by CPI Group (UK) Ltd, Croydon, CR0 4YY
01/10/2026
02226215-0001